NON GAMING PC Build Guide

A Primer for First Time PC Builders

Craig Shields

Non Gaming PC Build Guide
A Primer for First Time PC Builders

A Clock Press Book copyright © 2024
by Craig Shields
Published by Clock Press

Email inquiries: info@nongamingpc.org

ISBN: 978-0-9846718-7-8

Disclaimer
It is not the purpose of this guide to reprint all the information that is otherwise available to DIY computer builders, but to complement, amplify and supplement other sources. This book does not attempt to cover all possible computer builds.

This book is an independent guide created for educational purposes and is not affiliated with, endorsed by, or sponsored by any of the brands, manufacturers, or vendors mentioned within. All product names, logos, and brands are the property of their respective owners. Any references to specific products or brands are provided solely for illustrative purposes and do not constitute endorsements or recommendations.

Every effort has been made to make this book as complete and as accurate as possible. However, there may be mistakes both typographical and in content. Therefore, this text should be used only as a general guide and in combination with other texts.

The purpose of this book is to educate and entertain. The author and publisher shall have neither liability nor responsibility to any person or entity with respect to any loss or damage caused or alleged to be caused directly or indirectly by the information contained in this book. The reader is encouraged to use common sense, exercise patience and practice safe PC building practices at all times.

A Clock Press Book
www.clockpress.com

Contents

Introduction

Building a computer can seem like a difficult task, especially if you've never done it before. But like any project, it comes with its fair share of surprises, and sometimes, even the experts face unexpected problems. I remember this clearly from my time at Gateway 2000 in the mid-1990s during the famous Intel Pentium FDIV bug. It was a time when customers were alarmed over a flaw in Intel's early Pentium processors that caused certain math calculations to produce incorrect results.

At Gateway, we received a lot of customer calls about the issue, and there was plenty of anxiety surrounding it. The media called it a crisis. One of my coworkers, trying to keep things in perspective, reassured worried customers with humor, telling them, "They don't explode!" While the bug didn't cause any physical damage, the public reaction was serious enough to lead Intel to recall the affected CPUs.

This story from the past still feels relevant today. Both Intel and AMD, the two biggest players in the processor market, have recently experienced high-profile issues of their own. These challenges remind us that even the most cutting-edge technology isn't immune to setbacks. But they also remind us of something important: no matter the complexity of the technology, solutions are eventually found, products improve, and we keep moving forward.

The world of PC hardware is always evolving, and the key to staying ahead is understanding the basics. In this book, I'll guide you through building your own Non Gaming PC, with a focus on making sure you know the parts you need, what those parts do, and how they all work and fit together. After all, even if your build hits a snag, as I learned back in the Gateway days, there's always a way to get things running smoothly—and no, it won't explode!

The Non Gaming PC

Contents

Why build your Non Gaming PC? There are many prebuilt computers available. They come with comprehensive warranties, support options, and they're bursting with all the latest technology, so why put in all the work? Here are several reasons why you'll benefit from building your own Non Gaming PC:

Value: More bang for your buck with hand picked high quality components vs. generic parts.
Customization: You choose the right parts for you, and skip what you don't need.
Learning: Know all the parts, know what they do, and know how to replace/upgrade them.
Longevity: Compatible, high quality parts last longer.

We'll go into detail on these reasons in the following sections, but first let's answer the question of what is a Non Gaming PC?

1.1 The Non Gaming PC Defined

The Non Gaming PC gets your work done. It does the things you need a computer to do. It's a tool for your productivity. It may not have fancy lights or an all glass display case. It may not have an expensive graphics card and a 1200 watt power supply (unless you have a very specific need for those). It has only what you need for the things you need to do.

By contrast, the majority of prebuilt computers sold today are way overpowered for what they end up doing. Prebuilt systems often come loaded with high-end components such as powerful processors, large amounts of RAM, and even dedicated graphics cards intended for gaming or professional content creation. However, for the typical user, someone who spends most of their time browsing the web, checking email, working on documents, and streaming movies or videos, these advanced specs go largely unused.

The two most important attributes of a Non Gaming PC are reliability and usability. Reliability comes down to selecting high quality parts from well known manufacturers, with specifications supporting the fast and efficient execution of everything you'll do on your freshly built computer. Usability is all about the user interface, and it will define the experience of using your computer. The quality and reliability of the monitor, keyboard, mouse, and operating system will have a big impact on your satisfaction and day to day enjoyment of using your Non Gaming PC.

Another feature of a Non Gaming PC build is its focus on aesthetics. Unlike many gaming oriented systems, where RGB lighting and glass-panel "showcase" enclosures are prevalent, the Non Gaming PC build emphasizes function and simplicity over flash. In a Non Gaming PC build, the goal is a clean, efficient workspace rather than a visual statement.

Non Gaming PC style is defined by builds in understated cases, with practical airflow, noise reduction, and cable management in mind. The emphasis is on durability, ease of maintenance, and a clutter-free appearance. Components that prioritize silent operation, subtle designs, and efficient air cooling systems are preferred, creating a computer that blends into the home or office environment rather than standing out with bold lighting or flashy designs.

When you build your own Non Gaming PC with the help of this guide, you'll learn how to select the components you need. You'll know how the PC works, you'll know how to take it apart, and you'll know how to put it back together. You'll know what it's capable of, and what to upgrade if your needs change in the future. And you'll be happy with the looks of your new system.

1.2 Non Gaming PC Use Cases

The following are example use cases and configurations for comparison. You can always adjust what you need, these are just a baseline. In terms of budgeting, the next to last hardware generation will usually provide good value and performance. Right now, this is happening with PCIe 4/5 and DDR 4/5. Products in the 4th iteration of these technologies are less expensive than the latest 5th gen, but still give solid performance. As you'll learn later in this guide, PCIe is backward compatible. A motherboard supporting PCIe 5 SSDs will run PCIe 4 drives just fine, and you'll be able to upgrade to PCIe 5 devices later when the price comes down. When PCIe 6 comes out, the same thing will apply.

Three use case categories are defined below: Essential, Creator, and Producer. These categories serve as a rough guide to builds with escalating performance levels. These categories will appear in the parts section of this guide for reference when buying components. Throughout this guide, I will refer to the build categories as 'Essential' (**E**), 'Creator' (**C**), and 'Producer' (**P**) to streamline the discussion. For example, '**E/C/P**' refers to all three categories.

Essential systems are designed for everyday use, focusing on tasks like web browsing, office productivity, and media streaming. They offer reliable performance at a reasonable cost, making them ideal for those who need a solid machine for basic activities without requiring high-end hardware. The specifications are designed to let you know what to look for.

Web Browsing (social media, shopping, basic research)

Email and Communication (email clients, video calls like Zoom)

Office Productivity (word processing, spreadsheets, presentations)

Streaming Media (video/music streaming services like Netflix, YouTube, Spotify)

Basic Photo Editing (light editing, storing and organizing personal photos)

Online Banking and Shopping

CPU Specs: 3.3–4.3 GHz/ 4–6 Cores/ 8–12 Threads/ 8–18 MB L3 Cache/ iGPU

RAM: 16GB DDR4, (2x8GB), 3200–3600 MT/s

Storage: 500 GB PCIe Gen 3 or Gen 4 M.2 NVMe SSD

GPU: Integrated graphics only (sufficient for everyday tasks)

Creator systems are for those who engage in more demanding tasks such as photo and video editing, graphic design, or light 3D rendering. These PCs provide higher performance with upgraded components like additional RAM and more powerful processors, making them well-suited for creative professionals or hobbyists who need to handle multimedia tasks.

Photo and Video Editing (Adobe Photoshop, Premiere Pro)

Audio Production (podcast recording, basic mixing)

Graphic Design (vector design, infographics)

Light 3D Modeling or CAD Work

Advanced Multimedia Streaming (home media server, 4K video streaming)

CPU Specs: 3.5–5.1 GHz/ 6–10 Cores/ 12–16 Threads/ 20–64 MB L3 Cache/ iGPU

RAM: 32GB DDR5, (2x16GB), 4800–6000 MT/s

Storage: 1+ TB PCIe Gen 4 M.2 NVMe SSD for O/S & Apps, additional M.2 NVMe SSDs for larger files, HDDs for long term storage and backup

GPU: 12–16GB dedicated graphics card for smooth editing and rendering

Producer systems represent the highest tier, designed for power users requiring substantial computing power. These systems are optimized for complex tasks like software development, advanced data analysis, virtual machine use, or 3D modeling and CAD work. A Producer build prioritizes maximum performance and multitasking capabilities, with premium hardware that can handle intensive workloads and future expansion.

Software Development (coding environments, compiling)
Data Analysis (Excel with large datasets, Power BI)
Virtual Machines (running multiple OS environments)
Advanced Photo and Video Editing (4K video editing, complex renders)
3D Modeling and CAD (AutoCAD, SolidWorks)
Machine learning/AI
> CPU Specs: 3.0–5.8 GHz/ 12–24 Cores/ 24–32 Threads/ 30–128 MB L3 Cache/ iGPU
> RAM: 64GB DDR5, (2x32GB), 4800–6000 MT/s
> Storage: 1+ TB PCIe Gen 5 M.2 NVMe SSD for O/S & Apps, additional M.2 NVMe SSDs for larger files, HDDs for long term storage and backup
> GPU: 24GB dedicated graphics card for 3D rendering, virtual machines, and high-res media editing.

1.3 Value in building your Non Gaming PC

Depending on your budget, it is possible to save thousands of dollars by purchasing components and building your own system vs. buying a prebuilt. Prebuilt systems with the best performance and the latest processor (CPU), graphics card (GPU), and storage (NVME SSD) options are often priced much higher than the sum of the cost of their individual components. The below table is a comparison of a possible DIY build to a prebuilt system from a large, well known PC manufacturer, where the cost of parts plus the operating system software is considerably less than the purchase price of the prebuilt unit with similar specifications.

Potential Savings of Building your own PC

	DIY Build Estimate**		Well Known Brand Prebuilt
CPU	i7-12700K + Air Cooler	275	i7-12700K + Water Cooler
Graphics	Gigabyte RTX 3060 12gb vram	300	XX RTX 3060 12gb vram
Power Supply	Corsair RM750E 750 watt	100	800 watt proprietary
Motherboard	Gigabyte B760M	130	proprietary
Case	Corsair 3000D	70	proprietary
Memory	Corsair 32gb DDR5	95	32gb DDR4
Fans	Antec five pack, pwm	25	proprietary
Data Storage	2TB Western Digital	120	2TB Western Digital
Operating System	Windows 11 Home	140	Win 11 Home w/proprietary apps
		$1,255	$2,380

**Please note that the prices mentioned for components and pre-built systems are estimated market values intended solely for illustrative purposes. They are not offers to buy or sell and should not be taken as such. The technology market is highly dynamic, with prices fluctuating due to factors such as new product releases, demand, and supply chain conditions. We recommend conducting your own research or consulting with a professional to obtain current pricing information before making any purchase decisions.

Our estimated DIY build would cost $1,125 less than the manufactured unit, and it boasts identical or even better specifications. The DIY build features an effective and low maintenance air cooler for the CPU instead of water cooling, and even includes an upgrade to DDR5 memory technology. This kind of cost difference is not isolated to only a few systems or brands. Most prebuilt PCs cost considerably more than what it would cost to build a similar system. Some are a little closer than this example, some are much farther apart. As the equipment specs increase, the cost gap between a DIY build and a prebuilt tends to open up. As time goes on, prices will change, but the differences between the cost of DIY and prebuilt systems are likely to remain constant. Cost wise, the main strength of the DIY option is lower cost for better systems.

This is not to say that computer manufacturers do not provide value for that extra up front cost, they do provide considerable value. But that value is built into the purchase price. Manufacturers offer warranty options, product technical support and customer service, driver downloads, and updates to their proprietary system software, among other benefits.

What is one case where a prebuilt computer makes more sense from a cost perspective? At the low end of the price spectrum. Manufacturers can afford to build low end systems very inexpensively with bulk discounts on hardware and software. They also enjoy great economies of scale as they make and repurpose interchangeable components across various consumer, industrial, and international markets and brands. The below table is an estimate of a DIY build where the cost of parts plus the operating system software is more than the purchase price of a prebuilt computer with similar specifications. These systems do not include a display:

Low Cost DIY Build Estimate

	DIY Build Estimate, Low Cost**		Well Known Brand Prebuilt, Low Cost
CPU	AMD Ryzen3 5300G	80	AMD Ryzen3 5300G
Graphics	Integrated AMD Radeon Graphics		Integrated AMD Radeon Graphics
Power Supply	Thermaltake 500 watt	40	310 watt proprietary
Motherboard	Gigabyte A520M	70	proprietary
Case	Zalman T6 ATX Mid Tower	50	proprietary
Memory	Crucial 8gb kit (4x2) DDR4 3200	26	8gb (4x2) DDR4 3200
Fans, Mouse, Kybd	3 - 120mm Fans, USB Kybd/M	31	Fans, USB Kybd/M
Data Storage	256gb NVME, 1TB HDD, DVD-Writer	70	256gb NVME, 1TB HDD, DVD-Writer
Operating System	Windows 11 Home	140	Win 11 Home w/proprietary apps
		$507	$480

As you can see in the example above, large PC manufacturers can offer complete low end computers, with a licensed copy of Windows, for less than the retail price of all the components. Even the cheapest retail parts (not always the highest quality) will cost more at retail than a manufacturer will pay wholesale. Add in a licensed copy of Windows to match, plus the value of your time investment, and now your low end DIY build cost meets or exceeds the prebuilt retail price. Plus, the prebuilt system includes a warranty, technical support, and all the assembly work is done for you. The value that can be realized with higher end DIY builds doesn't exist at the low end of the price spectrum. When the absolute lowest total investment in time and money is the priority, going with the lowest cost prebuilt system can make sense.

1.4 Customization
What do you want your PC to do? The ability to select specific components, optimize performance for specific tasks, and tailor the looks to your taste are just a few of the benefits when building your own Non Gaming PC.

During your parts selection process, you can choose each component based on your specific needs and preferences. The choice of CPU, graphics card, motherboard, or a specific type of storage like M.2 NVME SSD, allows you to optimize the hardware to do the tasks you have in mind.

1.5 Learning
Building your first PC is a learning experience. What you're selecting parts, you'll learn the acronyms and terminology that refer to the different parts of the computer and the hardware technologies that are involved. With that knowledge you'll begin to learn about compatibility and how to select components and their various versions that will work together. During the physical build process, you'll learn about connectors and cable types, drives, cards, and slots, and how to

identify them by shape and function. You'll learn how to change settings in the BIOS, the chip that controls the basic input output systems on the motherboard. You'll be exposed to all of this during the PC building process, and it's a great learning opportunity.

1.6 Longevity and Future Usability

During your PC build process is the best time to select components that will last for years and be compatible with future iterations of your configuration. Cases, power supplies, storage devices, and expansion cards can often be swapped onto a new system board with the latest CPU years after your initial purchase. Careful parts selection and planning during the build can ensure the longevity of your computer and maintain the value of your time spent and money invested. DIY builders have access to retail components that are made with industry design standards in mind, and these retail parts are designed for flexibility and are intended to be interchangeable among any number of custom configurations. Retail PC parts have more value and a longer lifespan than proprietary parts contained in prebuilt systems.

1.7 PC Parts, Marketing, and Branding

When building a Non Gaming PC, the focus is on practicality, performance, and getting the most out of your investment without the extras designed for gamers. However, when you're selecting parts, you may come across components, especially motherboards, graphics cards, or even power supplies, that are branded with the word "gaming."

This doesn't mean those parts are unsuitable for a Non Gaming PC build. In fact, many "gaming" parts offer excellent specifications for tasks like productivity, media creation, or general use. After all, gaming can be one of the most demanding and resource intensive personal computer applications.

The big takeaway here is to focus on the features that matter to you and your build. If a "gaming" motherboard has the right number of ports, supports your CPU and RAM, and meets your budget, then there's no reason to dismiss it just because of the label. Marketing terms can sometimes be more about targeting an audience rather than defining the actual capabilities of the hardware.

Sometimes, parts with the right specs that have RGB lighting built in may be your only choice (and it can always be disabled later), and that's OK. In this case, look past the label and the appearance. Evaluate what the part offers in terms of functionality. The "gaming" designation might only affect some screen printed graphics, colors, lights, or a few features designed specifically for gaming, while the core specs could still be right for your build. Of course, if you have time you can always wait until the part with the right aesthetics and specifications becomes available.

Don't be surprised if you find yourself purchasing a part with the word "gaming" attached to it. The right hardware for your build may come from a variety of sources, and the best decisions come from understanding the specifications, not just the marketing.

1.8 Summary Judgement: Build Your Non Gaming PC

I'll end this section with a short summary of the advantages of building a Non Gaming PC over buying a prebuilt system. While prebuilt computers may seem convenient, they often come with additional costs. Support services from large manufacturers have free options but may require subscriptions for enhanced or priority levels, adding to the overall price. In contrast, building a Non Gaming PC can often be more cost-effective for similar specifications.

Then there's customization. Prebuilt systems come in predefined configurations, meaning upgrading a single component (like a graphics card) may force you to buy an entire higher-tier system with features you don't need. Building your own Non Gaming PC allows you to choose only the components that matter to you.

Prebuilt computers are designed to save manufacturing costs. Many prebuilt PCs use proprietary parts, such as cases, motherboards, and power supplies, that are not reusable in future builds or upgrades. These components are often designed specifically for the manufacturer's system and are not compatible with standard PC parts.

Manufacturers do the work of loading the Windows operating system on machines they sell. But they also do so much more. Prebuilt systems are often loaded with unnecessary software (bloatware) from vendors and partners, ranging from support tools to trial versions of subscription-based software. These programs can slow down the system by using memory and storage, and are generally designed to get you to spend more money. Removing this software is often time-consuming and frustrating.

1.9 Bottom Line

Building your own Non Gaming PC provides value, customization, learning opportunities and you end up with a long lasting, high quality computer with the specifications and aesthetics that you design.

I know this is a lot of new information to take in, but don't worry. In the next section, we'll cover all the parts, what they do, and explain the acronyms and abbreviations.

Summary 1: The Non Gaming PC

Benefits of Building your own Non Gaming PC

Value: High quality components of your choosing.

Customization: Select only the high performance options you need. Don't need a GPU? Skip it!

Learning: Know all the parts, know what they do, and know how to replace & upgrade them. Be the expert!

Longevity: Compatible, high quality parts last longer and upgrade well over their lifespan.

Reliability: Quality parts from well known manufacturers with specs that support your needs.

Usability: A great user interface sets the stage for satisfaction and day-to-day enjoyment of your Non Gaming PC build.

Aesthetics: Function and simplicity over visual effects. A clean and efficient workspace fits your environment.

Drawbacks of Prebuilt Systems

Cost: Prebuilt PCs can cost more than building a similar system. Higher specs mean better value for DIY building over prebuilt.

Compatibility: Prebuilts use non-standard, proprietary parts. Cases, motherboards, and power supplies often cannot be used in future builds or upgrades.

Bloatware: Unnecessary software from vendors and partners like support tools and trial versions of subscription-based apps. Bloatware take time and effort to remove, wastes resources, slows your PC, and wants you to spend more money.

Hidden Costs: Support services from large manufacturers may require subscriptions for priority support.

Low quality parts: Prebuilts often use inexpensive, lower quality connectors, fittings and other parts to save on manufacturing costs.

Parts Guide

Contents

2.1 Case and Fans

Computer cases protect components from damage and heat. The metal case shields the fragile components from physical damage while providing focused airflow to remove heat from the enclosed space. When selecting a case for your Non Gaming PC build, there are several factors to consider: size, design, aesthetics, and airflow. The size of the case will determine where it can fit in your space, and the size will also dictate some choices when it comes to internal components. Design features such as air filters, external drive bays (or the lack thereof), and front panel ports, will allow for some choices in the style and functionality of your PC.

Case Sizes

Computer cases come in all kinds of shapes, sizes, and designs, but they are often referred to by the size of the motherboard, the primary circuit board of the system, that the case can accommodate. So, selecting a case size also means you're selecting a motherboard size. The most common motherboard size is ATX, which stands for Advanced Technology Extended (patented in 1995 by Intel). The ATX specification covers the size of the motherboard, the mounting points where the board is attached to the case with screws, the location of the input/output connector panel where you attach your cables, and the size of the power supply and its power connection to the wall, among other things. When you select an ATX type case, that means the case will hold an ATX size motherboard and an ATX size power supply. ATX cases come in tower or mid-tower sizes. Tower means that the case is designed to stand upright. For example, the Corsair 4000D case shown in this guide is about 18" tall, 17" deep, and 9" wide. Another mid-tower slightly larger case from the same brand is about 20" tall. The full tower case is about 23" tall. An ATX size system is what I recommend to start with. An ATX tower or mid-tower size case is large enough to allow plenty of room to work, and is commonly available at reasonable prices and in a wide variety of colors and designs.

While compact cases designed for Micro-ATX and Mini-ITX motherboards might be appealing, they are not recommended for first-time builders. The limited space inside these cases makes assembly and component installation more difficult, and visibility is reduced. They may also require specialized cables or hardware modifications, adding unnecessary complexity to an already unfamiliar process. These factors can create a frustrating experience for a first build.

Case Design Features and Aesthetics

The design of the computer case offers some flexibility in building the kind of computer that looks appealing to you. Many ATX cases come with open panels and removable air filters. Some cases come with multiple glass panels. The case shown in this guide comes with a glass side panel, but Corsair also offers a solid metal panel that can be purchased as an add-on.

Front panel connections are very convenient. Consider where your PC will be located and think about what kinds of devices you would like to attach to it. Will you be using wired headphones? Then a case with an appropriate audio connector on the front panel would be in order. Cases generally have at least one USB port, a USB-C connector, and an audio jack on the front panel.

The once-common 5.25" drive bay used for CD or DVD drives has largely disappeared from modern cases due to the rise of downloadable software and streaming services. Many cases now use the front panel for airflow. Only a few cases still include front-facing external drive bays. If you occasionally need disc-based media, consider using a USB-connected external drive instead.

RGB LED Lighting

RGB LED Lighting (Red, Green, Blue Light Emitting Diode) is available, and while it isn't essential for a Non Gaming PC build, it may come built-in as part of certain components, and it can be used creatively. In a corporate or professional setting, RGB can be used to add visual flair to a presentation environment or workspace, especially in situations where aesthetics matter, such as client-facing desks or showroom setups. While it can be attractive, RGB lighting adds a layer of complexity to your build, and requires its own network of cables and connectors within the case. These cables are connected to the motherboard or a separate controller, which then interfaces with software for lighting control. There is also ARGB, which is addressable RGB. ARGB allows for more control over individual LEDs, enabling more complex lighting effects compared to standard RGB. While it is not my intention to cover RGB in-depth in this guide, I can offer one suggestion. If you are interested in RGB lighting, my suggestion is that you choose one vendor for all of your RGB components, to simplify the setup and control of your lighting. There is no one, industry wide, vendor neutral, agreed upon standard for RGB, so the best that you can do is stick with one vendor for all RGB hardware and software components to ensure compatibility and smooth operation.

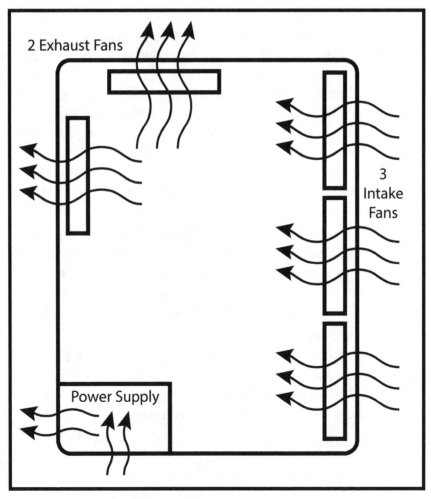

Positive Pressure Airflow

Fans

Computer cases need fans to move cool air into the case and exhaust heat. Fans come in several sizes, the most common are 120mm or 140mm, square. The typical mid-tower case will accommodate 3-120mm fans mounted vertically in the front panel of the case. Fans in the front

Fans expel air through the framed side.

of the case are commonly configured for intake of fresh air. There's a place for another fan at the back, setup as exhaust. And there is usually room for 2 or 3 fans at the top of the case. The front, top, and bottom panels of the case should be filtered, to trap large dust and debris before it makes its way into the case. You'll need a minimum of 3 fans in an Essential level Non Gaming PC build.

Computer case ventilation works best in a positive pressure configuration. To establish positive pressure, configure more fans as fresh air intake than exhaust (see left). The build shown here has 5 case fans: 3 in the front panel for air intake and 2 exhaust fans, one in the back of the case, the other exhaust at the top, above the CPU cooler, which has its own fans. With 3 intake fans and 2 exhaust fans forming positive pressure within the enclosure, the case won't draw dust and debris in through cracks and crevices as it would in a negative pressure configuration.

There is another fan located in the power supply. It draws air in from underneath the computer through the filtered panel on the bottom of the case, then exhausts through the rear of the power supply. Since the power supply doesn't affect the air inside the case in this configuration, it isn't counted when setting up positive pressure airflow.

Ventilated case panels should be filtered. If the bottom of the case is ventilated, make sure the case design allows for clearance between the vent and the floor or desk where your computer sits. Try to avoid placing the computer on carpet, but if you do, make sure there is adequate room for the vent to breathe.

To configure a case fan as intake or exhaust, notice how the fan is constructed. One side is open to the fan blade. The other side has frame supports covering the fan blade. A normally configured fan expels air through the framed side, while drawing air in from the open side. Some, but not all fans, have an arrow marked on the edge of the housing to show the direction of airflow. To install a fan, simply attach the fan to the case with the frame on the inside for intake, or place the frame to the outside for exhaust. When installing fans, always orient the fan housing so that the cable lead is closest to the motherboard mounting tray to ensure a short cable run and simplify cable management.

PWM (4 pin) Fans vs. 3 Pin Fans

When selecting fans you'll notice two types: PWM (pulse width modulation) 4-pin, and 3-pin fans. When the motherboard temperature sensors detect heat in the case or at the CPU, the system tells the fans to spin faster to bring in cool air and expel the heat. PWM fans allow the computer more control over the speed of the fan, allowing more control over cooling and fan noise. PWM controls the fan speed by varying the length of the pulses of power sent to the fan over the 4th pin, from about 20% up to 100% of the fan's rated speed. The 3 pin fans can only receive speed control by varying the voltage to the fan, which limits the high and low speed that they can attain. By fine-tuning the fan speed, PWM can decrease fan noise, especially under lower load conditions where full fan speed isn't necessary. Overall, the minor extra expense of PWM fans is worth the small price premium for a Creator or Producer level build. In an Essential level build, the 3 pin fans are perfectly fine.

Summary 2.1: Case & Fans

Recommended Configuration

ATX Case mid-tower size, for ample working room and compatibility. Recommended for all build levels **E/C/P**.

120mm fans, at least 3 for an Essential level build. Some fans will come with the case, buy extras if needed. PWM is optional.

More Intake: Create a positive pressure environment by having more intake than exhaust to minimize accumulating dust and maximize cooling.

Airflow: Fans expel air from the framed side. Circulate the air, in from the front, and out the top and back of the case.

Technologies Supported

Front Panel: USB, mic/headphones, power and reset switch.

Removable Filters: They trap dust and are easily cleaned.

Common Dimensions

Case: ATX Mid-Tower	**Fans:**
Height: 18" / 457mm	Standard is 120mm or 140mm
Length: 17.5" / 444mm	square and will usually fit 3-120's or
Width: 9" / 230mm	2-140's along the sides of the case.

Additional Features to Consider

CD/DVD-RW: External, if desired, when the front panel doesn't have a 5.25" bay.

RGB Lighting: If desired, get all the RGB lighting components from a single vendor if possible.

Warranty Period

Look for a 2 year warranty on cases and fans.

2.2 Motherboard

The motherboard is the backbone of the Non Gaming PC. It connects all the parts of the computer so they can be used by the operating system to do the things you want to get done. This section will tell you what you need to know when selecting a board for your build. I'll start with a review of the size of the board and the central feature of the motherboard, the CPU socket. Next, I introduce the motherboard chipset, which (with help from the CPU) really is the do-all workhorse of the system board that allows all the PC parts to work together. Then on to memory slots, USB, M.2, and SATA, voltage regulation, and some other nice-to-have features that indicate a quality board. At the end, I'll summarize everything and narrow it down to make your selection process easier.

Motherboard Form Factor and CPU Socket

The example build features an ATX size case, as discussed in the Case and Fans section, so I'll be selecting an ATX form factor motherboard to match.

The motherboard you buy will have a socket for the CPU, either AMD or Intel. The product description for the board will specify exactly what kind of socket is on the board. The AMD AM5 socket and the Intel LGA1700 socket are the most recent compatible sockets for currently available CPUs. These sockets have a frame and lever design where the lever is lifted, the CPU is simply laid onto the socket, and then the lever is closed and latched down, and the frame around the socket holds the CPU in place.

Chipsets Overview

The chipset, also known as the Platform Controller Hub (PCH), ensures that all the ports, slots, and hardware connected to the motherboard function seamlessly with the CPU. Each chipset is designed for a specific processor brand, AMD or Intel, and determines the types of CPUs and connectivity options a motherboard supports. Chipsets in a series can be roughly categorized as entry-level (AMD A and Intel B), midrange (AMD B and Intel H), or high-end (AMD X and Intel Z), aligning with the Essential, Creator, and Producer builds described earlier in this book. Let's explore how AMD and Intel chipsets match up with these build levels.

Essential Level: AMD A-series and Intel B-series chipsets

For an Essential-level Non Gaming PC build, which focuses on everyday tasks like web browsing, office productivity, media streaming, and light photo editing, both AMD's A-series and Intel's B-series chipsets offer budget-friendly options.

The AMD A-series chipsets, such as the A620, support PCIe 4.0 and provide sufficient connectivity options for basic needs. These chipsets do not support CPU overclocking, but this is not an issue for essential use cases where performance demands are moderate. The A-series is ideal for those looking to build an economical system with reliable performance, paired with a mid-range processor and integrated graphics.

The Intel B-series chipsets are similar, offering up to 20 PCIe lanes, and no support for CPU overclocking. While limited to USB 3.0 and below, these chipsets still provide adequate performance for essential tasks such as email, Zoom calls, and basic photo editing. Paired with a mid-range Intel processor, 16GB of RAM, and a 512GB M.2 NVMe SSD, B-series systems ensure smooth daily operations in an affordable price range.

Creator Level: AMD B-series and Intel H-series / Intel Z-series chipsets.
Creator-level builds are aimed at users who engage in more demanding tasks like photo and video editing, graphic design, and light 3D rendering. For these tasks, AMD's B-series and Intel's H-series chipsets strike a balance between performance and affordability.

The AMD B-series chipsets, such as the B650E and B650, offer PCIe 5.0 support for storage and optional support for graphics cards, making them well-suited for creative workflows that require handling large media files. With overclocking capabilities and up to 36 total PCIe lanes, the B-series allows for flexible system configurations, including multiple NVMe SSDs and dedicated GPUs. This chipset is a solid choice for those running creative applications like Adobe Photoshop or Premiere Pro, enabling smooth editing and rendering without overspending on hardware.

The Intel H-series chipsets are another strong option for Creator-level builds. Although they do not support CPU overclocking and offer fewer (up to 20) PCIe lanes compared to higher-end chipsets, they still provide enough connectivity for creative workflows. Paired with a multi-core Intel processor, 32GB of RAM, and a 12–16 GB GPU, an H-series system can efficiently handle the larger file sizes and complex operations typical of creative tasks like graphic design and video production. While it lacks some of the high-speed expansion options of Intel's Z-series, the H-series provides an excellent balance of performance and cost for creatives.

Differences to note: AMD's B-series chipsets offer overclocking and slightly more PCIe lanes than Intel's H-series, making them more suitable for users looking to maximize performance in creative applications. Bear in mind that Intel's Z-series chipset, while typically used in higher-end systems, could also be a good choice for a Creator-level build.

Producer Level: AMD X-series and Intel Z-series chipsets.
Producer-level builds cater to power users who need maximum performance for tasks like software development, advanced data analysis, virtual machine usage, 3D modeling, and machine learning. For these high-end systems, AMD's X-series and Intel's Z-series chipsets provide the necessary horsepower and flexibility.

The AMD X-series chipsets, including the X670E and X670, are the top-of-the-line enthusiast-grade options. These chipsets offer full PCIe 5.0 support for both storage and graphics, enabling high speed for data-intensive tasks. With up to 44 total PCIe lanes and overclocking capabilities, the X-series provides ample room for expansion and customization, making it ideal for complex workloads involving multiple GPUs, SSDs, and high-end processors. Paired with at least 64GB of RAM and a 24GB GPU, an X-series-based Producer build can easily handle the most demanding tasks such as 3D rendering, software compilation, and machine learning.

The Intel Z-series chipsets are Intel's high-end offering for Producer builds. These chipsets support CPU overclocking for "K" designated processors, allowing builders to push their hardware to its limits. With up to 48 total PCIe lanes and good USB support, the Z-series offers extensive expansion options, including multiple GPUs and NVMe SSDs. A Z-series-based system paired with a high-performance multi-core processor, 64GB or more of RAM, and a powerful 24GB GPU can easily handle intensive multitasking, virtual machines, and complex data analysis.

Motherboard Logical Diagram
On the next page is a generic, simplified, logical diagram of a typical CPU and chipset layout showing how the various parts of the motherboard are logically connected. The two main sections are the CPU at the top, and the chipset or platform controller hub (PCH), below.
The CPU is directly connected to the PCIe 5.0 bus and the PCIe 5.0 x16 slot, intended for the GPU card. The x16 is for the maximum number of PCIe lanes this slot can use.

PCIe is an expansion bus interface standard for PC components, most commonly storage devices and GPU cards. Each successive version of PCIe is twice as fast as the previous version. It is forward and backward compatible, so that a PCIe-4 device will work in a PCIe-5 bus slot, but only at the version 4 specified speed. Likewise, a PCIe-5 device will work in a PCIe-4 slot, but only at the version 4 speed. Unlike PCIe devices, DDR memory is NOT backward or forward compatible.

Motherboard Logical Diagram

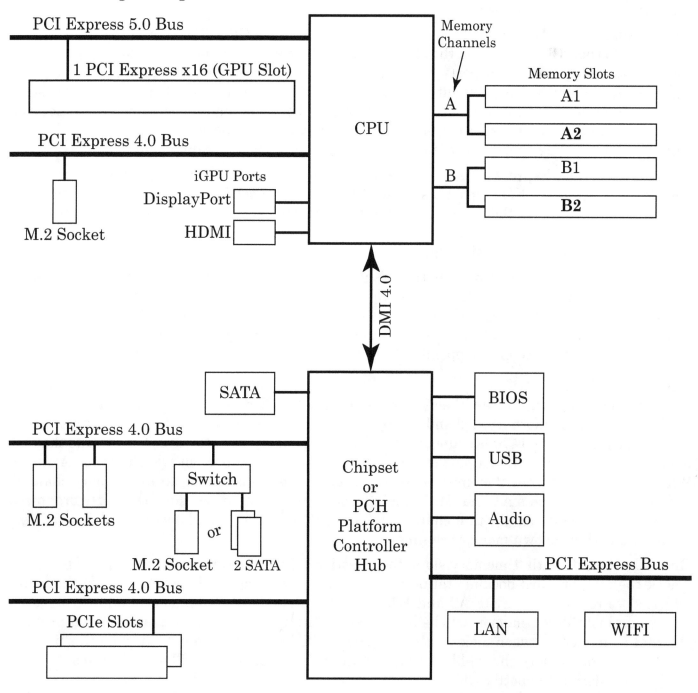

The DDR memory slots are directly connected to the CPU, where the memory controller is located. The CPU and motherboard together will dictate which memory architecture, DDR4 or DDR5, is supported by the motherboard.

The CPU is also connected to a portion of the PCIe 4.0 bus which is linking at least one M.2 socket directly to the CPU, which is where your first M.2 NVMe SSD drive should be installed.

The video connectors on the motherboard I/O panel, DP and HDMI, are for the integrated video (iGPU) on the processor and are directly connected to the CPU.

The CPU is connected to the Platform Controller Hub or chipset by Direct Media Interface or DMI 4.0. This kind of connection is very similar to a PCI Express connection using either 4 or 8 lanes, depending on the chipset. Even though there are fewer lanes in this connection, most of the devices connected via the chipset do not operate fast enough or use enough PCIe bandwidth to saturate the DMI connection.

Most motherboards have one M.2 slot connected directly to the CPU. The remaining M.2 slots are connected to the PCIe 4.0 bus via the chipset. In the diagram you'll notice a switch, where either one M.2 socket or two SATA ports can be used. This is a shared connection and is a common trade-off on system boards, consult your motherboard manual to see which of your ports are affected. If possible, use the other, unshared M.2 slots before using the switched slot. When you use it, take note of which SATA ports will become unavailable. There will be a number of SATA ports available on the board that are not affected by this sharing arrangement.

The rest of the PCI Express slots are connected via the chipset. USB, audio, ethernet, and WiFi are also connected via the chipset, as is the BIOS. As discussed in the chipsets overview, the chipset controls most of the features on the motherboard, and more advanced versions of the chipset support a wider variety of connections, system control options via the BIOS, and faster connections than the more basic chipsets.

A note on BIOS: Most modern motherboards use UEFI (Unified Extensible Firmware Interface), which has replaced the older BIOS (Basic Input Output System), to manage their configuration. However, manufacturers still refer to this utility as the BIOS. For simplicity, we'll refer to the motherboard configuration utility as the BIOS in the rest of this guide. If you need to update the BIOS on your system, refer to Appendix A at the end of the Build Guide for an overview.

DDR RAM Slots

Most motherboards will have either 2 or 4 memory slots, allowing various configurations based on the design of the motherboard and the capabilities of the CPU. For instance, consumer-grade CPUs, like the Intel i7-14700K, support dual-channel memory architecture. This means that the memory controller, which is a part of the CPU, can manage memory in two channels, A and B. When a motherboard has 2 memory slots, each slot corresponds to a separate memory channel, commonly referred to as A1 and B1. In motherboards with 4 memory slots, these are organized into pairs (A1 and A2 for the first channel, B1 and B2 for the second), allowing each pair of slots to operate within its own memory channel.

On motherboards with 4 memory slots, the question is, where to put only 2 sticks of memory? Check the motherboard documentation to be sure. It's often marked right on the motherboard which slots to use first. Slots A2 and B2, the bottom end of each memory channel, are used first when installing 2 memory sticks in a 4 slot motherboard. This is because when a memory channel is divided between 2 slots, it's designed to fill from the bottom to the top. In this case, think of a 2 slot memory channel like a cup for water. It's impossible to fill only the top of the cup, the water must fill the bottom first.

Will 2 sticks work right next to each other on one channel, using slots A1 and A2? No. It will run poorly, if at all, and it won't run in dual channel mode. Placing 2 memory sticks on one channel cripples the performance of the memory controller by placing all the load on one channel instead of balancing the load between the 2 available memory channels. This is why motherboard manufacturers specify which slots to use when using only 2 sticks of RAM in a 4 slot board.

RAM slots A2 and B2 identified.

What about using slots A1 and B1? No. This leaves the bottom end of the memory channels vacant, leaving an empty space on the memory circuit for signals to echo and slosh around, creating problems for the memory controller.

What about just one module?
Only use pairs of memory modules in your build. Using a single RAM stick ignores all the advantages of dual channel memory architecture. Instead of one 32GB stick of RAM, use 2-16GB modules to balance the load across both channels of the memory controller.

What about using all 4 memory slots? Yes, this can work (see also page 35), but you may need to disable any memory overclocking functions like *XMP/EXPO/DOCP, and run the memory at a slower speed to get a stable system when all 4 memory slots are filled. After filling all 4 slots, confirm that the system boots and check the BIOS settings to see if memory overclocking is still enabled and what speed the memory is using. If it won't boot, or if you start having crashes or memory errors, disable XMP. The system will probably then run perfectly fine with the 4 memory modules running at a slightly slower speed. Why is this a potential problem? Because the memory controller has to work twice as hard to manage both channels filled with 2 RAM modules each. Depending on how the CPU and the memory modules fared in the silicon lottery, the memory controller, RAM, and motherboard together may or may not be able to run the system in a stable state with all 4 memory slots filled while running at XMP rated speeds. Think of the XMP rated memory speed like a car speedometer. The fact that the speedometer goes up to 160 mph is not a guarantee that the car will go that fast, only that the gauge could display that speed under the right conditions.

Selecting a motherboard with 2 or 4 memory slots comes down to your intended purpose for the system. If you want or need the ability to install the most RAM possible, even at a slightly slower speed, say for video editing, virtualization or complex graphic design, get a motherboard with 4 memory slots. Motherboards with 4 memory slots are also the most commonly available product. Even if you're only planning to get one pair of memory modules, having a 4 slot motherboard will give you an opportunity to increase your installed memory later on.

*XMP = Extreme Memory Profiles by Intel | EXPO = Extended Profiles for Overclocking by AMD | DOCP = Direct Overclocking Profile by ASUS-AMD

USB

Universal Serial Bus (USB) connectors come in a variety of configurations. Most motherboards will have USB ports on the rear I/O panel, to be accessed from the back of the PC case, and onboard USB connectors to connect to the USB ports on the front of the PC case. Cable connectors on the motherboard are often referred to as headers. Header is short for "pin header" and means that the connection will be made by pushing the cable connector down over a row of pins. These are usually keyed so that they can only go on one way and they should be clearly marked.

The motherboard in this guide has 10 USB ports on its rear I/O panel, and 4 USB headers internally on the motherboard. There are several different connection speed specifications between the available connectors. The tables below provide details on USB port types, how they are named and the data transfer speeds of the listed USB ports.

When using USB ports, save the higher speed ports for data transfer devices, like USB flash drives, Micro SD card readers, external optical drives, scanners, or printers if they can take advantage of the higher speed. Unless otherwise specified, a mouse, keyboard, drawing tablet or other input device can be plugged into one of the USB 2.0 ports. Also, note that the internal motherboard USB-C connector in the example table only supports 10 Gbps (gigabits per second), so the PC case mounted USB-C port will be limited to the 10 Gbps speed. If you have a device

Motherboard Rear Panel USB

Typical Motherboard Rear I/O Panel USB Connectors			
Port Type	**Description**	**Data Rate**	**Signaling Rate**
USB 4	USB-C only	up to 5 GBps	up to 40 Gbps
USB Type-C USB 3.2 Gen 2x2	The x2 indicates there are 2-10Gbps lanes combined to achieve 20Gb/s.	up to 2.5 GBps	up to 20 Gbps
USB Type-A (red) USB 3.2 Gen 2	Also known as USB 3.2 Gen 2x1. The x1 equals one 10Gb/s lane.	up to 1.25 GBps	up to 10 Gbps
USB Type-A (blue) USB 3.2 Gen 1	Also known as USB 3.0, USB 3.1 Gen 1, and USB 3.2 Gen 1x1.	up to 625 MBps	up to 5 Gbps
USB Type-A (black) USB 2.0/1.1	Basic USB port.	up to 60 MBps	up to 480 Mbps

1 byte (B) = 8 bits (b)
Signaling rate refers to the speed at which data is transferred over the USB connection.
Data rate is the actual throughput, considering protocol overheads and other factors.

Mbps: Megabits per second
MBps: Megabytes per second
Gbps: Gigabits per second
GBps: Gigabytes per second

Motherboard Internal USB

Typical Motherboard Internal USB Headers and Connectors			
Port Type	**Description**	**Data Rate**	**Signaling Rate**
USB Type-C socket USB 3.2 Gen 2	Also known as USB Type-E.	up to 1.25 GBps	up to 10 Gbps
USB 3.2 Gen 1 header	19 pins.	up to 625 MBps	up to 5 Gbps
USB Type-A USB 3.2 Gen 1	Also known as USB 3.0, USB 3.1 Gen 1, and USB 3.2 Gen 1x1.	up to 625 MBps	up to 5 Gbps
USB 2.0/1.1 header	9 pins, basic USB port.	up to 60 MBps	up to 480 Mbps

that can use the full USB 3.2 Gen 2x2 20 Gbps speed, you'll have to plug it into the back panel of the motherboard to take full advantage of that connection. Another thing to be aware of is that, while USB-C connectors are capable of supporting high speed connections, they are not always the fastest port on the computer. USB-C connector speeds start at 5 Gbps and go up from there. Be sure to check the specifications of the port, and the device you're connecting, to determine what connection speeds are supported by both devices.

M.2 Slots and SATA Ports

When planning your PC build you'll want to know the number and types of storage devices you can attach to your system. The motherboard will have multiple interfaces for storage devices and this will dictate the number of drives your system can accommodate and how they are connected.

M.2 connector slots and SATA (Serial AT Attachment) ports are the most commonly available storage interface connections available on consumer motherboards. M.2 is both newer and faster technology than SATA. The motherboard shown in this guide has 4-M.2 connector slots on the PCIe 4.0 bus and 6-SATA connectors. Please refer to the motherboard diagram for a typical logical arrangement of M.2 and SATA connectors.

Use the M.2 slot closest to the CPU for an M.2 NVMe SSD drive, and install your operating system and applications on that drive. Your data, applications, and operating system will benefit from the speed that M.2 NVMe storage provides.

Voltage Regulator Module (VRM)

The VRM on a motherboard manages power delivery from the PSU to the CPU and other key components. It functions by stepping down the voltage received from the PSU to the lower voltages required by the CPU, RAM, and other integrated circuits. A high-quality VRM and heatsink ensures stable and efficient power delivery, which is essential for the reliable performance of these sensitive components, especially during high loads. When selecting a motherboard, consider the quality and design of the VRM, because a robust VRM setup enhances system stability and longevity, and works in tandem with the PSU to provide consistent power.

Other nice-to-have features

Audio support: In the past, a dedicated sound card was required to convert digital audio signals to analog for playback on speakers or headphones. Now, audio support chips are built-into most motherboards, the most popular audio chipset being made by Realtek. The Realtek Audio CODEC (coder/decoder) supports high definition audio in 2, 4, 5.1, and 7.1 channel configurations.

Video support: This guide recommends getting a CPU with a built-in graphics processing unit (iGPU) for backup and troubleshooting. The motherboard in this guide supports onboard graphics with HDMI and DisplayPort connectors on the rear panel. To use the onboard video, simply plug your monitor cable into the port on the back of the motherboard, no extra GPU card required.

Wired and WiFi networking: Most motherboards will include at least one RJ45 jack for a wired ethernet connection. WiFi networking is an available option as well. The WiFi network antenna connectors will be located on the motherboard rear I/O panel. This is another function that used to require an additional expansion card that is now built-in to some motherboards.

Bluetooth: If you have or plan to get any Bluetooth accessories like headphones or speakers, be sure that the motherboard you select supports Bluetooth. It will be listed under wireless communication in the motherboard manual.

7-Segment Display: A display mounted on the motherboard that shows numeric error codes and boot codes in case something goes wrong, simplifying troubleshooting when boot errors occur. The motherboard shown in this guide has 4 status LEDs for CPU, DRAM, VGA, and BOOT that light up if the device isn't working properly. These lights are OK and many systems have them, but a 7-segment numeric display can be much more informative, which will make troubleshooting a boot issue faster and easier.

Boot Speaker: The majority of motherboards no longer include a built-in PC speaker or buzzer. To regain this useful feature, you can add a boot speaker. This small device helps you troubleshoot startup issues by providing beep error codes during boot if there's a problem. The speaker emits a single chirp during boot to indicate a successful system start. Instructions for adding a PC boot speaker are included in the build section.

Last but not least: Download and read a copy of the manual before purchasing a motherboard to make sure it fully supports all the features you are interested in.

Boot Speaker

Summary 2.2: Motherboard

Essential (**E**), Creator (**C**), Producer (**P**)

Recommended Configuration

ATX Size Motherboard that matches your CPU and Socket

> **Essential:** AMD A-series | Intel B-series Chipset
> **Creator:** AMD B-series | Intel H / Z-series Chipset
> **Producer:** AMD X-series | Intel Z-series Chipset

RAM Slots: 4 for future memory upgrades. **E/C/P**
M.2 Slots: At least 2 for more for increased primary storage capacity and upgrades. **E/C/P**
SATA Ports: At least 4 or more, depending on how many additional storage devices you need. **E/C/P**

Additional Features to Consider

7-Segment Display: Would be a nice feature to have for troubleshooting boot issues.

Warranty Period

Look for a 3 year warranty on the motherboard you select.

Technologies Supported

Connectivity Options: USB, SATA, PCIe x16, HDMI & DP video, Audio, RJ45 Gigabit Ethernet, WiFi, Bluetooth.

2.3 Central Processing Unit & Cooler

Choose the processor for your build. The choice of processor will influence the compatibility and performance of other components. Here's a hardware-level introduction: The processor is a small, squarish board with a silicon chip package mounted on top, hidden under a metal heat spreader known as the integrated heat spreader (IHS). On the underside of the processor, you'll find metal contacts, either pins for a Pin Grid Array (PGA) socket or flat pads for a Land Grid Array (LGA) socket. The processor socket is located on the motherboard, which was discussed in the last section. The rest of this section will be a brief overview of CPU manufacturers, fabrication and sorting processes, and the concept of the 'silicon lottery.' Then to help you select a CPU, I'll outline the specifications, what they mean, and what to look for including cores & threads, clock speed, cache size, cooling & TDP, integrated graphics (iGPU), chip sockets, locked vs. unlocked processors, and pairing the CPU with an appropriate GPU.

CPU Manufacturers

The two primary CPU companies are Advanced Micro Devices (AMD) and Intel Corporation (Intel). Both design their own processors.

However, the actual fabrication of these chips takes place in semiconductor fabrication plants, known as fabs. Intel manufactures the majority of its chips in its own fabs. By contrast, AMD is a fabless semiconductor company, similar to Apple and Nvidia, meaning it designs its processors, but outsources the fabrication to Taiwan Semiconductor Manufacturing (TSMC), the world's largest chipmaker.

CPU Fabrication and Sorting

While making computer chips, such as CPUs or memory, is an extraordinarily complex process, it is not perfect. The combination of minute imperfections in processes and materials results in chips with varying speeds and capabilities. These discrepancies have become increasingly narrow over time. Fabricated chips are tested and then sorted into performance-graded 'bins' based on their test results. From these 'bins', the various performance grades of chips are assigned. A chip that fails testing at the highest speed may qualify to be sold as a slower product. Similarly, a chip where some cores fail testing might have those cores disabled and be sold as a lower-grade product with fewer cores.

Manufacturers aim to meet specific yield goals, and chips may be altered post-manufacturing to meet these targets. For example, consider a scenario where a batch of chips all emerge perfectly, no defects, and all qualify for the highest possible grade. Some of these perfect chips could be deliberately altered to fit the specifications of lower grades to meet production.

The previously mentioned sorting 'bins' are directly connected to overclocking and the 'silicon lottery.' In the past, the process of testing processors and selecting bins was less precise, often resulting in CPU chips that exceeded the performance specifications for their assigned bin. If an end user could manually increase the clock speed and raise the CPU voltage beyond known specifications while maintaining stability, they were said to have won the 'silicon lottery'. Over time, efficiency improvements in silicon manufacturing have reduced the available headroom for overclocking, and the 'silicon lottery' has largely lost its significance. This has made CPU overclocking much less beneficial for most users, particularly those not looking to push their systems to extreme limits. Memory overclocking is still relevant; it is necessary to enable *XMP/ EXPO/DOCP (Extreme Memory Profile) in the BIOS to achieve the advertised memory speed for your build.

When selecting a CPU for your build, consider how you intend to use your computer. Use the Essential, Creator, and Producer level Non Gaming PC builds described in Chapter 1 as a guideline. Here's a brief recap of the CPU specs in each build for reference while you're learning about processors in this chapter:

Build Level	Clock Speed (GHz)	Cores	Threads	L3 Cache (MB)	Video Capable
Essential	3.3–4.3	4–6	8–12	8–18	iGPU
Creator	3.5–5.1	6–10	12–16	20–64	iGPU
Producer	3.0–5.8	12–24	24–32	30–128	iGPU

CPU Cores & Threads

The featured build in this guide has an Intel Core i7-14700K (14th gen) processor. It has 20 cores (8 performance at up to 5.6GHz, 12 efficient at up to 4.3GHz) running 28 threads. Each of the performance cores can run 2 threads, and the efficient cores run one thread each. Each core is a physical processor in the CPU. In this processor, the performance cores are designed to handle demanding tasks, such as video editing, by operating at a higher clock speed of up to 5.6GHz. On the other hand, the efficient cores, running at up to 4.3GHz, are optimized for less intensive tasks and power efficiency, which is good for background processes and everyday applications like email or web browsing. This combination allows for a balance of high performance and efficiency.

A thread is a stream of instructions that can be processed by a core. With hyper-threading technology, as seen in the performance cores of the Intel Core i7-14700K, a single core can handle two threads at once, doubling the number of tasks it can process. This leads to improved multitasking in applications designed to take advantage of multiple threads.

Clock Speed

The clock speed of the CPU sets the overall pace for processing in the computer. For the featured build, the max turbo frequency of the performance cores is 5.6 GHz. That is 5.6 billion clock cycles per second, billions of simple mathematical and logical operations, that perform the work of running a computer program. But clock speed isn't as big a factor as it appears. This is because the architecture of the CPU influences how much work is accomplished in each clock cycle. Things like cores, threads, cache size, and the arithmetic logic unit (ALU), all affect system performance. In fact, some processors can do more work per cycle at a slower clock speed than other faster

*XMP = Extreme Memory Profiles by Intel | EXPO = Extended Profiles for Overclocking by AMD | DOCP = Direct Overclocking Profile by ASUS-AMD

CPUs, due to the efficiency of their architecture. So, the clock speed, while it does matter, must be considered in combination with the other features of the CPU when making a buying decision.

Cache Size

The cache in a CPU stores copies of data and program instructions from RAM that are frequently accessed by the CPU. This reduces waiting time and bottlenecks, because the CPU can quickly retrieve data from the cache rather than waiting for it to be copied from much slower RAM.

Cache memory, or static random access memory (SRAM), is faster and has lower latency compared to dynamic random access memory (DRAM). SRAM is constructed with six transistors per bit of storage, making it physically larger than DRAM, which uses one transistor and one capacitor per bit. The need for a charge cycle in DRAM's capacitors introduces a delay, making SRAM the much faster technology.

Cache is organized into levels - L1, L2, and L3. The L1 cache, being the smallest and fastest, is the first place the CPU looks for data. Located closest to the CPU cores, L1 provides rapid access to data. If data is not found in L1 (a 'cache miss'), the CPU then searches the larger and slightly slower L2 cache. The L3 cache, while the largest and slowest among the three, is still much faster than retrieving data from RAM. It holds information not contained in L1 or L2 caches.

For example, the Intel i7-14700K processor features a cache arrangement of about 1.8MB in L1, 28MB in L2, and 33MB of L3 cache. This substantial amount of cache helps in quickly accessing data and instructions. The higher performance CPUs will generally have more cache.

When choosing a CPU, consider the amount of cache memory. A larger cache can contribute to the processor's performance, especially in tasks that require rapid access to large amounts of data or complex instructions.

CPU Cooling & Thermal Design Power

There are two primary CPU cooling methods: air or liquid. Both use a cold plate attached to the CPU with thermal compound to absorb heat. Air coolers use heat pipes and a fin-covered heat sink to disperse heat, while liquid coolers circulate fluid through a radiator to remove heat. This guide's build uses an air cooler for its simplicity and durability, with installation details provided in the build section. Air coolers, though larger and potentially interfering with other components, have fewer moving parts and easily replaceable fans. Liquid coolers may offer superior cooling but involve more maintenance and have more failure points over time due to liquid permeation.

CPU Air Cooler

Thermal Design Power (TDP) is a term frequently encountered when selecting CPU coolers. You'll see TDP mentioned in CPU specifications in different ways. Sometimes it's called thermal design power, or thermal design point. Intel calls it "Processor Base Power". It's often described as a measure of the maximum amount of heat a CPU is expected to generate under typical load. However, TDP is a manufacturer provided guideline and is not a precise indicator of cooling requirements. Different CPU manufacturers calculate TDP using varying methods, leading to inconsistencies. Cooler manufacturers might use different criteria for determining the TDP ratings of their products. This disparity means that a cooler's TDP rating might not always line up perfectly with the TDP rating of a CPU. While TDP can provide a general idea about a CPU's heat output, it shouldn't be the sole factor in choosing a cooler.

With the issues surrounding TDP, thorough research becomes essential when selecting a CPU cooler. One way is to consult the recommendations provided by cooler manufacturers. Many of these manufacturers test their coolers with a variety of CPUs and often list compatible models on their websites, offering a more reliable guide than TDP ratings alone. Exploring reviews and benchmarks from trusted tech websites and user forums is another way to see how well a cooler performs in real-world scenarios.

Beyond TDP ratings and manufacturer recommendations, several other factors should influence your cooler selection. Consider the physical size of the cooler and whether it fits in your case, especially if you're working with limited space. Confirm that the cooler comes with a mounting kit that is compatible with your CPU's mounting socket. Also, take into account the overall airflow design of your system – a cooler's efficiency can be greatly affected by the case's air intake and exhaust capabilities (see the case and fans section of this guide). The climate where you live, and your typical usage patterns also play a role in determining the cooling capacity you'll need.

If you're doing everything right in terms of case ventilation and cooler installation, and you purchase a well tested and high quality cooler, and you're taking your specific circumstances into account, either kind of cooler, liquid or air, will meet your CPU cooling needs.

Integrated Graphics Processing Unit (iGPU)

An integrated graphics processing unit (iGPU) is a graphics rendering chip built-into a CPU. It handles the basic graphics functions of a computer, eliminating the need for a separate GPU card. Confirm the presence of an iGPU by checking the specifications and description of your chosen CPU. All Intel Core series and the new Core Ultra 200S series chips contain an iGPU except those marked with an F suffix in the part number, which indicates that an iGPU is not present. With AMD chips, Ryzen chips with a G in the model number will contain an iGPU. In any case, always refer to the product description to confirm the existence of an iGPU in the processor you are considering.

You'll want to select a processor that has an iGPU. While an on chip video solution will not match the processing power of a GPU card in the areas of graphics processing or professional video editing, there are several reasons to get an iGPU:

- Backup and troubleshooting: If your GPU fails or is otherwise unavailable, the iGPU will allow you to use the system normally until your GPU is working.
- Energy savings: Using an iGPU will cut down on the amount of electricity your computer uses. If you're not using any graphics intensive applications, an iGPU is a good choice.
- Heat: GPU cards produce heat, much more so than an iGPU. Choosing to stick with the iGPU will reduce the amount of heat inside the PC case.
- Multiple monitors: If all the monitor ports on your GPU card are full and your system has an iGPU, you can plug one more monitor into the motherboard graphics connector to take advantage of the iGPU and use the basic graphics capabilities on that monitor.
- Small form factor computers: If, in the future, you decide to build a smaller PC, an iGPU will provide graphics functionality with zero extra room required inside the case.

Overall, the cost differential is pretty small between a CPU that has an iGPU and one that does not. Because of its utility and low cost, including an iGPU in your build is a good idea and adds value to your finished system.

CPU Sockets

A processor is connected to the rest of the computer in a special purpose socket on the motherboard specifically designed to fit that type of CPU. Consumer grade motherboards have just one CPU socket. There are multiple kinds of CPU socket, be sure to note which socket type your CPU is compatible with.

AMD and Intel employ a variety of socket types in their CPU designs. Motherboards and their supporting circuitry are built to support either an AMD or an Intel CPU. While AMD and Intel chips sometimes share a socket type, such as Land Grid Array (LGA), the design, pinout, and circuits of the sockets are exclusive to their brands and are not interchangeable.

The build shown in this guide uses an Intel Core series i7-14700K CPU, which is compatible with the LGA 1700 socket. The LGA 1700 socket was introduced in 2021, and is compatible with Intel 12th, 13th, and 14th generation Core series CPUs. The LGA 1700 socket contains 1,700 pins that contact the base of the CPU which has flat pads or a 'land grid array' on its underside. The pins in an LGA socket are fragile, and they are protected by a cover removed during CPU installation.

The LGA 1700 socket replaces the LGA 1200 socket previously used by Intel. The LGA 1200 is a 1,200 pin socket, compatible with Intel 10th and 11th generation Core series processors.

Socket AM5 is the most recent socket (2022) from AMD; it is an LGA 1718, supporting the AMD Ryzen 7000 series of processors. The AM5 socket is the successor to the AMD AM4 socket. AM4 was introduced in 2016 and is a pin grid array (PGA) socket accepting 1,331 contact pins from AMD PGA chips in the Excavator, Zen, Zen+, Zen 2 and Zen 3 CPU families.

AMD Pin Grid Array (PGA) Processor

The choice of motherboard, CPU, and cooler go hand in hand. When selecting your motherboard, make sure it supports the CPU you are planning to get. Motherboards are clearly identified as to the supported CPU and socket type. Match the cooler with the CPU socket.

Locked vs. Unlocked Chips

Whether to buy a locked or unlocked chip is another choice to be made during the PC parts selection process. When a processor is sold with an unlocked clock multiplier, the advantage is that it can be "overclocked" to change its settings like voltage, and clock speed, to enhance performance. The advantages to be gained by overclocking are much less meaningful now than they used to be, as previously discussed in the CPU fabrication and sorting section, but this doesn't mean that an unlocked chip isn't a good buy. Even if you're not planning to actively overclock, an unlocked CPU can still offer several benefits:

1. Higher Base Performance: Unlocked CPUs often come with higher base clock speeds compared to their locked counterparts. This means they offer better performance right out of the box, without any manual overclocking.
2. Flexibility: Having an unlocked CPU means you have the option to overclock in the future if your needs or interests change.
3. Resale Value: Unlocked CPUs often retain their value better and are more attractive in the second-hand market, potentially making them easier to sell if you decide to upgrade later.

4. Higher Quality Silicon: Manufacturers often use higher quality silicon for their unlocked CPUs, which can translate to more stable performance and potentially longer lifespan.
5. Compatibility with High-Performance Motherboards: Unlocked CPUs are often paired with higher-end motherboards that offer better features, such as improved power delivery, more control over system settings in the BIOS, more connectivity options, and audio support. Even if you don't overclock, these motherboards can offer a better overall system experience.

AMD and Intel each handle overclocking in their own way. All AMD Ryzen processors are unlocked by default. AMD Ryzen 7000 series processors do not include a cooling heatsink or fan, and a CPU cooling solution must be purchased separately.

Intel processors in the Core series and the new Core Ultra 200S series will have a K suffix in the part number to indicate that they are unlocked. The processor shown in this guide, the Intel i7-14700K, is an unlocked chip. Intel does not supply a CPU cooler with their unlocked chips.

Intel's locked chips, those without the K suffix in the part number, come with a heatsink and fan in the box. The locked chips from Intel present a simple, drop-in solution as opposed to an unlocked CPU for which you'll need to purchase a cooling system.

CPU / GPU Pairing

When choosing a CPU for your Non Gaming PC, try to match it with your GPU. The CPU handles general tasks, while the GPU renders graphics. If the CPU is too weak for the GPU, a 'CPU bottleneck' can limit performance. For instance, a high-end GPU like the NVIDIA GeForce RTX 4090 paired with a low-end CPU may not reach its full potential. To prevent this, balance the CPU's power with the GPU's capabilities. High-end GPUs pair best with Intel Core i7/i9 or AMD Ryzen 7/9, while mid-range GPUs suit Intel Core i5 or AMD Ryzen 5 processors. This balance ensures optimal performance without one component limiting the other.

Summary 2.3: CPU & Cooler

Recommended Configuration

AMD or Intel CPU with iGPU: AMD, G suffix. Intel, no F suffix.

Build	GHz	Cores	Threads	L3 Cache (MB)
Essential	3.3–4.3	4–6	8–12	8–18
Creator	3.5–5.1	6–10	12–16	20–64
Producer	3.0–5.8	12–24	24–32	30–128

Energy Efficiency

TDP Rating: Check the TDP ratings for an idea of the cooling requirements and power draw. Higher watts/TDP is more heat.

Additional Features to Consider

Locked or Unlocked CPU: Get a locked Intel CPU with the included cooling fan solution for simplicity.
Get an unlocked AMD or Intel K series CPU to keep your future options open. AMD Ryzen CPUs are not locked.

Thermal Compound: Bonds the air cooler heat sink to the CPU integrated head spreader. Comes with the air cooler.

CPU Cooler: Air cooler with 2 fans, front to back airflow.

Warranty Period

A 3 year limited warranty for boxed processors is standard.

2.4 Random-Access Memory (RAM)

Random-Access memory is your PC's workspace. Fast, efficient, and appropriately sized memory will increase the speed and usability of your computer. There are some specific points to consider about memory when building your computer. I'll go over them here, and some of this material will be covered again in the build section of this guide.

When you are selecting memory for your Non Gaming PC build, there are some numbers to be aware of regarding the size, speed, and technology of the modules.

DDR5 Memory Module. Notice the offset location of the key notch on the edge connector.

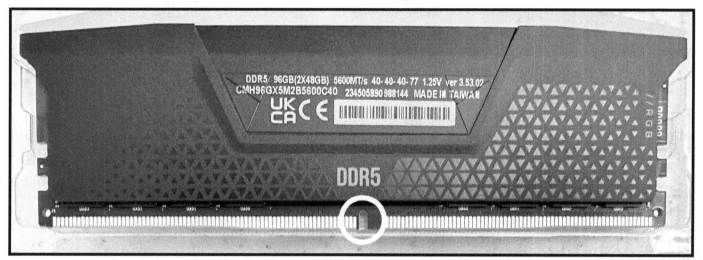

DDR5 96GB(2X48GB) 5600MT/s 40-40-40-77 1.25V

{
DDR5: Double Data Rate 5
96GB(2X48GB): Size of the memory module
5600MT/s: Speed in MegaTransfers per second
40-40-40-77: Latency timings (lower is faster)
1.25V: Operating voltage

Size

Memory size is measured in gigabytes (GB), which represent about 1,000 megabytes (MB) each. (One megabyte is about enough to hold the text of a 400 page book.) You'll need at least 4GB to run Windows 11, but this isn't really enough for most applications such as web browsers and spreadsheets to run smoothly. Adobe Photoshop requires 8GB of RAM minimum. Other productivity applications may require even more RAM. Also consider that running multiple applications at the same time can exhaust the amount of available memory. Memory modules are sold in kits of two modules, so a 32GB kit will consist of a matched pair of modules, each 16GB in size. If you're only using 2 modules, install them in slots A2 and B2 on your motherboard.

When installing 2 memory kits to fill all 4 slots at once (see also page 25), even if they are identical kits, keep each kit on its own channel to help ensure system stability. In this case, kit #1 goes in slots A1 and B1, kit #2 goes in slots A2 and B2, filling all 4 memory slots. It is not recommended to mix memory modules from different manufacturers in any configuration.

Speed

DDR (Double Data Rate) Memory commonly includes XMP (Intel), EXPO/DOCP (AMD) profiles to enable speeds beyond the base clock speed. Without these profiles enabled in the BIOS, the memory operates at its standard speed. As discussed in the motherboard section, *XMP/EXPO/ DOCP are considered to be overclocking, and while the speeds are not guaranteed, they are programmed into the chips and are considered to be stable by the manufacturers. Keep all this in mind when considering the cost of the fastest and most expensive memory modules.

DDR technology performs data transfers on both the rising and falling edges of each clock cycle, doubling the effective data rate compared to Single Data Rate (SDR) memory, which transfers data once per cycle. Memory speed is measured in MegaTransfers per second (MT/s), with each transfer occurring on either the rising or falling edge. Therefore, a 5600 MT/s rating implies 5600 million transfers per second, leading to an actual clock speed of 2800 MHz for DDR memory, due to the double data transfer mechanism (transfers happen twice per clock cycle).

Latency timings, often referred to as just "timings," are a measure of how many clock cycles it takes for RAM to respond to requests from the CPU. These timings affect the speed at which the RAM can read and write data. For these numbers, lower is better. The first number listed, and most often referenced, is the CL or CAS Latency, which is the time it takes, measured in clock cycles, between the memory controller requesting data and the data being available to be sent on the memory bus. The RAM modules with lower or "tighter" timings are often more expensive than other slightly slower products.

Your motherboard manual will detail the memory speeds supported by the board, depending on the CPU that is installed, since the memory controller resides within the CPU. The motherboard manufacturer will also publish a QVL or Qualified Vendor List detailing the memory modules, with speeds and part numbers, that have been verified and tested to work with your exact motherboard and BIOS revision. Be sure to consult the manufacturer's QVL and your motherboard manual when selecting memory for your build to ensure compatibility.

Technology

DDR4 and DDR5 are the most recent iterations of memory technology, with DDR5 being the latest. If your budget allows, and your CPU/Motherboard supports it, go ahead and spend the extra to get DDR5. DDR5 is faster, more energy efficient, and comes in larger capacity modules than DDR4. If budget considerations outweigh these advantages, DDR4 is a viable option and will certainly perform well in most configurations. As previously noted, your selection of CPU and motherboard will determine which technology will work in your system.

Faster memory is going to cost more. Consider what you're planning to use the PC for most of the time. If you're involved in time intensive projects like rendering videos where time is a factor, an investment in faster memory may make sense in the long term.

*XMP = Extreme Memory Profiles by Intel | EXPO = Extended Profiles for Overclocking by AMD | DOCP = Direct Overclocking Profile by ASUS-AMD

Summary 2.4: Random-Access Memory (RAM)

Recommended Configuration

Build	Size	Tech	Modules	Speed (MT/s)
Essential	16GB	DDR4	(2x8GB)	3200–3600
Creator	32GB	DDR5	(2x16GB)	4800–6400
Producer	64GB	DDR5	(2x32GB)	4800–6400

Warranty Period

Most memory manufacturers offer a limited lifetime warranty.

Additional Features to Consider

Latency Timings: Lower numbers are better but probably cost more and provide minimal overall benefit.

QVL: Check the Qualified Vendor List from your motherboard's manufacturer for memory module compatibility.

XMP/EXPO/DOCP: Enable memory overclocking in the BIOS after Windows is installed to get the advertised speed from your memory modules.

2.5 Storage

If **RAM is your PC's temporary workspace, then data drives are your PC's file cabinet.** Unlike RAM memory, which is volatile and loses all data when the computer is turned off, data storage devices use non-volatile memory to preserve your 'virtual file cabinet' of photos, emails, documents, and programs, even when your PC is powered down. These devices come in various shapes and sizes. In this section, I'll explore the different types of data storage devices and the specifications you need to know to select the right storage for your Non Gaming PC build.

Hard Disk Drive

The mechanical hard disk drive (HDD) is an old data storage technology that is still relevant today. HDDs work by encoding data in a non-volatile, shallow magnetic coating on a spinning non-magnetic platter, with a read/write head moving back and forth over the surface of the platter. For long-term storage of infrequently accessed data, the HDD is a good alternative. The drawbacks of HDD technology are slow data access speeds, sensitivity to physical shock, and noise created by the mechanical components. Some HDDs are faster than others, but keep in mind that the higher RPM drives, say 7200 RPM, will use more power and produce more noise. Drives in the 5400–5600 RPM range will use less power and run more quietly. The HDD shown in this guide uses a 5400 RPM drive with the SATA interface. Commonly available HDDs are 3.5" in size and include a Serial AT Attachment (SATA) version 3.0 interface. Hard drives generally transfer data at a rate of between 100 MB/s and 200 MB/s, and are nowhere near saturating the 600 MB/s data rate capability of

8 TB Hard Disk Drive

the SATA III interface. Select a HDD for your build if want to make local backups of your data for long-term storage. Don't plan on running your applications or operating system from a hard disk drive; the hardware is simply too slow compared to current technology. The next step up in data storage technology is the solid state drive (SSD).

Solid State Drives (SSD) and Non-Volatile Memory Express (NVMe)

SSDs use NAND flash memory to store data, replacing traditional mechanical hard drives (HDDs). They are silent, shock-resistant, have no moving parts, and are known for their high speed and low power consumption. The three most common types of SSD storage device for consumer PCs are: the 2.5" SATA SSD, the M.2 SATA SSD, and the M.2 PCIe NVM Express (NVMe) SSD.

2.5" SATA SSD: These drives are usually the least expensive per GB of the three SSD variants. 2.5" SSDs mount to the case in designated spots using screws and take up very little space. Because they are a lightweight plastic shell construction, they can also be mounted with double-sided tape in any convenient spot. They do require a SATA data cable and a power cable. Performance-wise, these drives are SATA III compatible and will move data at around 500 to 550 MB/s, which is near the theoretical limit (600 MB/s) imposed by the SATA III interface. In fact, some SATA SSDs are limited in their data transfer rate by the capabilities of the SATA interface. Overall, SATA SSDs provide less performance and are a poor value when compared to the newer technology of PCIe NVMe based devices.

M.2 SATA SSD: These drives connect to the motherboard via a SATA M.2 slot which has two key notches to accommodate the edge connector of the M.2 SATA drive. An M.2 NVMe PCIe SSD has only one key notch (see right). A SATA device connected via an M.2 slot still runs under the same 600 MB/s limitation as any other SATA III connection. This type of drive has the advantage of being smaller than a 2.5" SSD, which would be beneficial in a tight small form factor case or laptop application, but performance wise is equal to the 2.5" unit.

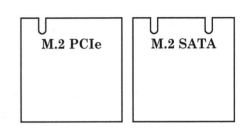

M.2 PCIe NVMe SSD: NVMe M.2 SSDs are the fastest storage option, using the NVMe protocol on PCIe lanes to achieve higher data transfer rates. Developed as a replacement for the older SATA interface, which was originally designed for spinning hard disk drives and couldn't fully exploit the speed of flash-memory-based SSDs, NVMe drives connect

M.2 PCIe NVMe SSD

directly to the system via the PCIe interface, providing wider and faster data paths compared to SATA. For example, a 4-lane PCIe Gen 3 NVMe SSD can reach speeds of around 4 GB/s, while PCIe Gen 4 SSDs can achieve up to 8 GB/s. PCIe 5.0 storage devices take this even further, offering up to 14-16 GB/s with the same 4-lane configuration. Gen 5 performance can be ideal for tasks requiring heavy data throughput like, Video editing, 3D modeling, Machine learning and AI development, and CAD/CAM applications.

For an Essential-level Non Gaming PC build, a PCIe Gen 3 or Gen 4 NVMe SSD provides ample performance, making them both great choices for everyday tasks like web browsing, office work, and media consumption. At the Creator level, a Gen 4 SSD is an ideal fit, delivering improved speed for more demanding workloads such as content creation and video editing. Due to the price premium, PCIe Gen 5 SSDs are best for the Producer level build, where high-performance applications justify the cost.

Information about the number of PCIe lanes (x4 or x2) and the specific generation of PCIe supported by the M.2 connectors on your motherboard should be detailed in your motherboard manual. In this build, I'll install the first NVMe drive in the M.2 slot next to the CPU, which has a slightly faster connection to the PCIe bus. This drive will be the optimal location for your Windows operating system and application files. Your data, such as photos, videos, and documents, can also be stored on this drive or on an additional NVMe drive.

NVMe Size: When you see the number **2280** on an NVMe drive, this is referring to the physical size of the memory module in mm. **'22'** is the standard width for the M.2 connector, while **'80'** refers to the length. The 2280 size is the most commonly available, but other sizes like 2230 and 22110 are also options. Most motherboards provide at least one slot with a support post at the 80mm length for NVMe drives. There are adapters available if you want to use a shorter 2230 drive in a 2280 slot.

DEFRAG is Not for SSDs: Defragmentation is a process designed for traditional hard disk drives (HDDs) where data is physically organized to be stored in contiguous sectors on the disk platters. This reduces the time it takes for the drive head to move to different tracks to read a whole file, thereby improving access times and read/write speeds on an HDD.

For Solid State Drives (SSDs), including all types of NVMe and SATA SSDs, defragmentation is not only unnecessary but can be detrimental for several reasons:

No mechanical movement: SSDs use NAND flash memory to store data, which can be accessed at almost uniform speed regardless of where the data is stored in the drive. There's no physical read/write head to move across a spinning platter, so there's no benefit in organizing files to be contiguous.

Wear and tear and the warranty: NAND flash memory has limited write cycles, so each cell can only be written a certain number of times. Processes like defragmentation can shorten its lifespan. Manufacturers warranties are usually 3 to 5 years, and include a Terabytes Written (TBW) rating, indicating the amount of data that can be safely written before risking data integrity.

Built-in wear leveling: SSDs have internal wear-leveling algorithms that intentionally spread out write and erase cycles across the memory cells to extend the life of the drive. Defragmentation interferes with these algorithms.

TRIM command: Modern SSDs and operating systems support the TRIM command, which helps to manage the SSD's performance and wear by marking blocks of data that are no longer in use as ready to be erased.

Don't run DEFRAG on an SSD. Windows 10/11 automatically recognizes SSDs and will not schedule automatic defragmentation for them. Instead, Windows will use optimization techniques like the TRIM command to help maintain the performance and longevity of the SSD.

Relative Size of Common Data Storage Measurements

8 bits = 1 byte (01010101)	Enough to store a single character of text, like the letter 'A' or '!'.
1000 bytes = 1 kilobyte (kb)	About one page of plain text, a short email or a small text file.
1000 kb = 1 Megabyte (MB)	Approximately a 1,000-page book or a 4-second high-fidelity audio file. This would take up the majority of space on an old style 3.5 inch floppy diskette.
1000 MB = 1 Gigabyte (GB)	Around 1,000 books of 1,000 pages each. For media, it's about 200 songs or a standard-definition movie. A GB could hold a few minutes of HD video.
1000 GB = 1 Terabyte (TB)	Equivalent to about 1 million books or 1,000 gigabytes. A TB can store around 250,000 songs, more than 300 hours of high-definition video, or roughly 100,000 high-quality photographs.

Summary 2.5: Storage

Note: The speeds listed are "up to" maximum theoretical values. Actual performance can vary based on the system's configuration and usage.

Recommended Configuration

Place an M.2 NVMe SSD in the slot next to the CPU: Install Windows and applications on this drive. This will be your **C:** drive. Add other data drives after Windows setup.

The most bang for the buck: The latest generation of PCIe storage will be the fastest, but also the most expensive. Select the previous generation (4 instead of 5) for the best value.

M.2 NVMe Drive Specifications

Build	Size	Technology	Speed
Essential	500 GB	Gen 3/4	4 GB/s
Creator	1 TB+	Gen 4	8 GB/s
Producer	1 TB+	Gen 5	16 GB/s

No Defragmentation for SSD Devices

Wear and Tear: The defragmentation process wears out NAND flash memory which has a limited lifespan measured by the number of terabytes written (TBW) to the device.

SSD Warranty Period

SSD manufacturers warrant their products by time, usually between 3–5 years, and by terabytes written (TBW).

Max Data Rates in Megabytes & Gigabytes/Sec

SATA HDD	**100–200 MB/s**
M.2 SATA SSD	**500–550 MB/s**
M.2 NVMe SSD PCIe Gen3 x4 Lanes	**4 GB/s**
M.2 NVMe SSD PCIe Gen4 x4 Lanes	**8 GB/s**
M.2 NVMe SSD PCIe Gen5 x4 Lanes	**16 GB/s**

2.6 Power Supply

The power supply is like the heart of the PC. Often referred to as a PSU (power supply unit), it converts alternating current (AC) from your local utility to direct current (DC) and safeguards against instability by regulating the output. Computers use DC because it only flows in one direction, making it easy to regulate for low-voltage computer components. The motherboard receives the DC power from the PSU, further regulates current through the VRM (see also page 27), converts voltages, and distributes power to the rest of the components.

In this section, I'll go over the basic electrical requirements for the PC, how to calculate the wattage to figure out what size PSU to get, features to look for, power supply efficiency ratings, and some thoughts on shopping by brand and warranty.

Basic Electrical

You'll need a convenient, grounded, electrical power outlet near your PC setup. Because electrical surge protection depends on proper grounding, make sure the power outlet you select is a modern 3-prong outlet and confirm that the grounding is set up properly with an inexpensive outlet tester. If the tester indicates a fault in the circuit, get an electrician to correct any issues before using that socket to power your build.

Ideally, the PC should be on its own dedicated circuit breaker. Avoid having any appliances on the same circuit as your PC. Appliances, like a washing machine, can cause voltage drops, power spikes and/or surges, emitting electrical noise into the circuit. You also want to avoid overloading the circuit. Household circuits have a maximum current capacity, generally 15 or 20 amps. Running a high-load appliance and a computer on the same circuit can easily exceed this capacity, leading to tripped circuit breakers or blown fuses. This can not only interrupt your work but also poses a safety hazard.

Once you have a known good power outlet, you'll need a surge protector power strip to plug in the power cables for your PC, monitor, and anything else directly attached to your computer. A quality surge strip will protect the PC from overcurrent, overload, and short circuit conditions. Some may even include filtering to reduce electrical noise and interference (EMI/RFI). Look for a surge protector with a high joule (energy absorption) rating and a Connected Equipment Warranty (CEW). While surge protectors defend against spikes, they don't provide backup power during outages. For complete protection, especially in areas with unreliable power, consider using an uninterruptible power supply (UPS) along with a surge protector.

Wattage Calculation

When considering what PSU to get for your Non Gaming PC build, I'll offer these guidelines up front: For an Essential level PC build with no GPU card, get a 600 Watt modular PSU which is 80 PLUS Gold and CYBENETICS certified. That will be more than sufficient for the life of that build. For Creator and Producer level Non Gaming PC builds, do a rough calculation of the power you'll need based on your hardware configuration, and add a 25 percent margin for safety, which is the approach I outline in the rest of this section.

To figure the wattage for the power supply for your PC build, start by listing all the components in your system, including the CPU, GPU, motherboard, RAM, storage drives, cooling fans, and anything else plugged into the motherboard or backplane. Next, find the power consumption for each component. For the CPU, check its Thermal Design Power (TDP), which shows the normal power consumption under load. Since GPUs are a power-hungry component, get the TDP of your GPU as well. Figure that the motherboard will use around 60 watts, give or take. Each stick of RAM will use about 5 watts, HDDs generally about 10 watts each, and SSDs use around 5 watts. Figure that cooling fans use around 4 watts each. If any of these components have LED lighting attached you can bump up the estimate slightly, check the package or manufacturer's website for more precise figures. It's better to err on the high side when calculating power usage. Once you have the power consumption for all components, add them together to get the total,

Component	Watts	Quantity	Total
CPU TDP	125	1	125
Motherboard	60	1	60
GPU TDP	165	1	165
RAM w/RGB	9	2	18
M.2 SSD	5	3	15
HDD	10	1	10
Fans	7	5	35
			428
Safety Margin			*25 percent*
Total Watts w/margin			535
Recommended PSU size			**600 / 650**
upgraded GPU w/margin	450		826
Recommended PSU size			**850 / 1000**

Power Supply Sizing

then add a safety margin of 20-30% to account for potential future upgrades, component aging, and to ensure the PSU operates efficiently with less strain. If you think you might upgrade the GPU within the foreseeable future, consider including the TPD for the upgraded GPU in your calculations now; that way you won't have to buy a new power supply when you upgrade the GPU later. The table shows the calculations for the build shown in this guide. There are also online power supply calculators which will yield slightly varying results but are still a handy and time saving resource.

Features to look for in a Power Supply
Modular Cables: Power supplies in the past were non-modular, designed so that all the cables were permanently attached to the unit and could not be detached or swapped out. This resulted in a clutter of unused wires and connectors, taking up space in an already crowded case. Modular PSUs allow you to connect only the cables you need, reducing clutter and improving airflow inside the case. This makes building and upgrading the PC easier and neater. Make sure the PSU you select has the right types and number of connectors for the system you're planning to build. One final note about modular power cables: Only use the cables that came with your power supply or are made specifically for your power supply model. **NEVER MIX AND MATCH POWER SUPPLY CABLES**. Connector pinouts and modular cable designs can contain subtle modifications between the previous and most recent version of the same PSU, not to mention different models. Even though the connectors will physically fit, the voltages on the pins may have changed, or the cable design may be different, leading to unexpected results. Using the wrong power supply cables can cause catastrophic damage to your computer. If you ever upgrade your PSU, discard all your old cables and replace them with the new cables from the new PSU.

ATX 3.0 / 3.1: The ATX 3.0 power supply specification introduces the 12VHPWR connector (also known as the PCIe 5.0 connector). It's a 12-pin design (12 main power pins and an additional 4 signal pins) that can deliver up to 600 watts of power. Designed specifically for powering GPUs, this connector replaces multiple 8-pin connectors, simplifying cable management. The ATX 3.1 specification makes improvements on the 12VHPWR connector (12V-2x6), and focuses on efficiency, stability, and safety in computer power supplies. When you have a choice, select a PSU with the most recent ATX specification which will include the latest improvements.

ATX Size: When selecting a power supply for your first build, choose one with the ATX form factor. This specification matches with the case and motherboard, ensuring the PSU fits properly and is compatible with your other components. An ATX power supply measures 150mm in width, 86mm in height, and around 140mm in depth. The depth can vary, which may limit the available space for cable routing in certain case designs. To address potential space constraints at the front of the power supply, Corsair's RMx Shift Series (shown in this guide) has the connectors relocated to the side, which can be more convenient for cable management.

Zero RPM Mode: This is a nice feature that increases the lifespan of the power supply. It turns the PSU fan off during low to moderate loads, which reduces noise and dust intake, decreases power usage, and saves wear and tear on the fan. The fan only activates when the thermal controls detect rising heat levels in the case.

Power Supply Efficiency Ratings

There are two primary certification programs available for computer power supplies: 80 Plus and CYBENETICS. The power supply used in the featured build, the Corsair RM1000x Shift, has both certifications, see the image at right. Here's an overview.

Close up photo of the Corsair RM1000x package showing efficiency certifications.

80 PLUS Certification Program: The 80 PLUS certification program, managed by CLEAResult, is a globally recognized standard for energy efficiency in computer power supplies. This certification measures the efficiency of a power supply at various loads (20%, 50%, and 100%) and ensures that it meets specific efficiency thresholds. The program has multiple tiers, including 80 PLUS, 80 PLUS Bronze, Silver, Gold, Platinum, and Titanium, each representing increasing efficiency levels. A higher certification indicates that the PSU wastes less power as heat, leading to lower electricity bills and reduced environmental impact. Their highest certification, the Titanium level, measures efficiency at a 10% load, a state where many computers spend a significant portion of their time.

CYBENETICS Certifications: CYBENETICS offers a comprehensive certification program that evaluates both the efficiency and noise levels of computer power supplies. Unlike the 80 PLUS program, which primarily focuses on efficiency, CYBENETICS assesses PSUs at multiple load points under realistic operating conditions, offering a more detailed view of performance. The efficiency ratings range from ETA-A+ (most efficient) to ETA-D (least efficient). Similarly, the noise level ratings range from LAMBDA-A+ (quietest) to LAMBDA-D (noisiest). This dual certification system helps consumers choose power supplies that not only save energy but also operate quietly, making them well suited for environments where noise is a concern.

Shopping by Brand and Warranty

When shopping for a power supply, it's a good idea to consider the brand, warranty, and the product's reputation in reviews and online forums.

Reputable brands like Corsair, Seasonic, and Cooler Master provide varying warranty periods that reflect their confidence in product durability and performance. Corsair offers warranties of up to 10 years on many of its power supplies. Seasonic, another well known brand, goes a step further with warranties extending up to 12 years. Cooler Master, while offering shorter warranties of up to 5 years, still provides solid and dependable power supplies suitable for a wide range of builds. However, many other brands also produce high-quality power supplies, so research and consider all available options because technology is constantly improving.

In addition to warranty coverage, consider product reviews at retailer sites and look at online forums where other users share their experiences. These reviews and discussions can provide valuable insights into the real-world performance and reliability of different power supplies, helping you make a more informed decision.

Considering the brand, warranty, and online reputation can help you choose a PSU that not only meets your performance needs but also ensures sustained support and security for your PC build.

Summary 2.6: Power Supply

Recommended Configuration

600W Minimum: Headroom and efficiency improve when there is more power to work with. Also consider the ability to upgrade and keep the same PSU.

ATX 3.1: Supports the 12V-2x6 specification for GPU power.

Surge Protector Power Strip: Choose a well known brand that offers a Connected Equipment Warranty (CEW) and has a high joule (energy absorption) rating.

Efficiency Certifications: A unit that has both the 80 Plus and CYBENETICS ratings has passed numerous tests from two different certifying programs. Look for 80 Plus Gold minimum.

Airflow: Isolate the PSU airflow from the rest of the system by placing the intake at the bottom of the case to draw cooler air from outside and exhaust it directly.

Common Dimensions

PSU Size: ATX **Height:** 86mm **Depth:** 140mm or more **Width:** 150mm	**Fan:** Most PSUs use a 120mm to 140mm square fan. A larger fan can move more air at a lower speed and create less noise.

Additional Features to Consider

Zero RPM Mode: Reduces noise to virtually silent during low to moderate loads and extends the lifespan of the fan by minimizing wear.

Fully Modular Cables: Fewer cables keeps your build neat and promotes air circulation. Never swap cables between power supplies. Only use the cables that came with your PSU.

Warranty Period

Look for a 5 year warranty at a minimum.

2.7 GPU & VRAM

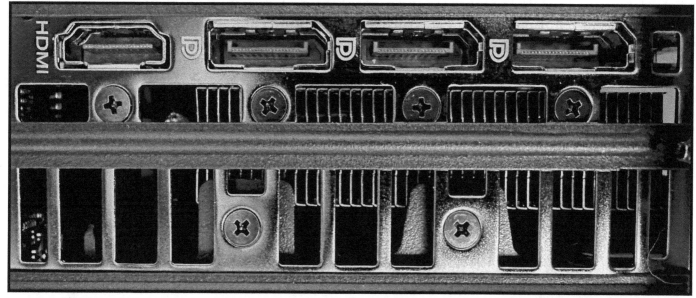

GPU card, 2 expansion slots wide, 1 HDMI port, 3 DP ports. Note the connector shapes.

The graphics processing unit creates the image you see on the screen. A powerful GPU enhances high-resolution visuals, accelerates video editing and 3D rendering for content creators, and boosts performance for tasks like working with local machine learning models and generative AI. It also improves everyday productivity, such as managing multiple displays and ensuring smooth video playback. Video random-access memory (VRAM) complements the GPU by enabling it to handle high-resolution textures, complex scenes, and multitasking more efficiently. In this section, we'll explore how the GPU and VRAM work together, and review GPU specifications to help you make an informed decision for your Non Gaming PC build.

GPU-VRAM vs CPU-RAM

The GPU and VRAM are like specialized vers ions of the PC's CPU and RAM. While the CPU and RAM are capable of handling all types of computing, including video processing, they do so at a slower pace compared to the GPU and VRAM. Here is an example to illustrate: Imagine two car washes in the same neighborhood: Car Wash A offers a full range of services, from washing and waxing to interior vacuuming and wheel polishing, but it has only one lane, processing cars one after another at a rate of 10 per hour. In contrast, Car Wash B specializes in fast exterior washes with 20 dedicated lanes, and each lane washes 20 cars per hour. After one hour, Car Wash A has completed 10 cars with full service, whereas Car Wash B has completed exterior washes on 400 cars. Similarly, the CPU and RAM (Car Wash A) can perform a wide variety of computing tasks, including video processing, but not as efficiently as the GPU and VRAM (Car Wash B), which are optimized for high-speed, parallel processing of graphics and other suitable work. This specialization makes the GPU/VRAM combination extremely effective for tasks that benefit from parallel processing, such as video rendering, editing graphics, or machine learning.

When choosing a GPU, there are several specifications and technologies where a little bit of knowledge can help you make an informed buying decision.

Shader Cores
These determine the GPU's ability to render graphics, perform complex calculations, and process visual effects in real-time. Shading, texturing, and lighting enhance visuals in applications like video editing, 3D modeling, and AI processing. The more shader cores a GPU has, the better it can handle detailed graphics and complex workloads.

Ray Tracing
Ray tracing is a rendering technique that simulates how light interacts with objects in a scene, creating realistic lighting, shadows, and reflections. If you want lifelike, cutting-edge graphics, support for ray tracing is a must-have feature for your GPU.

GDDR6 (VRAM)
High-speed memory is essential for handling large textures and ensuring smooth performance in video editing applications. Graphics Double Data Rate 6 is a type of high-speed memory that offers higher bandwidth and faster data transfer rates compared to its predecessors. GDDR6 memory ensures smooth and efficient handling of large datasets and complex graphics. Consider having at least 12 GB of VRAM in your GPU as a baseline. Go with 16GB or more if you want to run applications or workflows that can take advantage of the extra horsepower.

CUDA
CUDA (Compute Unified Device Architecture) is NVIDIA's parallel computing platform, allowing efficient processing of complex computing jobs, like machine learning. It enables applications to perform complex computations much faster than traditional CPU-based methods.

Stream Processors
Stream processors are the primary processing units in AMD GPUs, similar to NVIDIA's shader cores. They handle various tasks, including rendering graphics, performing calculations, and executing parallel processing operations. As with shader cores and CUDA cores, more stream processors means better overall performance for an AMD GPU.

Size
GPUs come in various sizes, from compact single-slot cards to larger triple-slot behemoths. When choosing a GPU, you need to ensure it fits in your PC case without obstructing access to motherboard features like M.2 NVMe slots or other nearby components. Check the length, height, and width of the GPU, as well as the number of expansion slots it occupies. Larger GPUs often require more power and have more fans for cooling, which may affect your case's airflow and power supply requirements. Always check your case's specifications for maximum GPU length and clearance. Most mid-tower ATX cases will have no problem handling 280–310 mm GPU.

Power Requirements
Consider the power requirements of the GPU when planning your Non Gaming PC build. High-performance GPUs demand a lot power, often necessitating a robust power supply to ensure stable operation. When selecting a GPU, check its power consumption, measured in watts, and ensure your PSU can deliver adequate power, including extra headroom for other components. A GPU will require specific power connectors from the PSU. Insufficient power can lead to system instability or failure to boot. A well-matched PSU not only supports the GPU but also contributes to the overall reliability and longevity of your PC. See the power supply section of this guide for more about choosing a PSU.

Ports

The video ports on the rear panel of the GPU determine how you connect your monitors and other display devices to your PC. When choosing a GPU, check the types and number of ports it offers to ensure compatibility with your monitors. DisplayPort (DP) and HDMI are the most widely used. DisplayPort is the preferred connection for most use cases because it supports higher resolutions and faster refresh rates than HDMI. If you want to run multiple monitors, confirm that the GPU you have in mind has enough ports of the correct type. Properly matching your GPU's video ports with your monitor setup ensures optimal display quality and functionality for your build.

Resolution Support

The resolution a GPU supports, such as 1080p (Full HD), 1440p (2K), and 4K, determines the clarity and detail of the images displayed on your screen. Higher resolutions like 4K offer more pixels and finer details, enhancing visual quality for professional graphic work. However, higher resolutions also require more graphical power to maintain smooth frame rates. When selecting a GPU, consider the resolution you plan to use in your workflow.

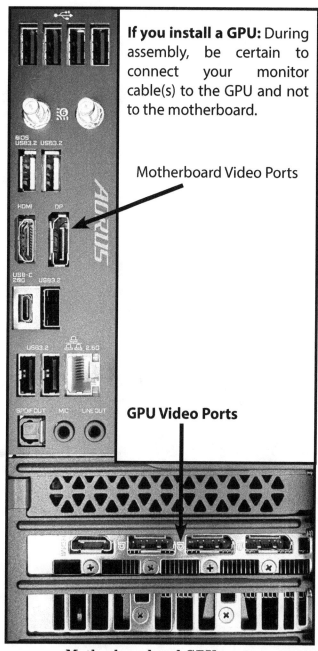

If you install a GPU: During assembly, be certain to connect your monitor cable(s) to the GPU and not to the motherboard.

Motherboard Video Ports

GPU Video Ports

Motherboard and GPU ports

Summary 2.7: GPU & VRAM

Recommended Configuration

12GB VRAM is a good starting point, 16 GB+ for faster throughput in graphics heavy workflows.

Connectivity: DP and HDMI ports should be available, confirm how many you need of each type.

Processing: Shader Cores, Ray Tracing, Stream Processors, and CUDA Cores are features to compare between GPUs.

Warranty Period

A 3–4 year warranty is commonly available.

Additional Features to Consider

Power: Refer to the power specifications for the card you are considering and select your power supply accordingly.

Resolution support: Confirm that the card is designed to be paired with your monitor - 1080p (HD), 1440p (2K). 4K.

Airflow: GPU fans push air into the card, expelling heated air into the computer case, through the sides and slightly through the back of the card. A blower style card expels heated air directly out of the back of the case through the expansion slot vents.

2.8 Monitor, Keyboard & Mouse

The display is the primary interface for your PC. Choosing the right monitor will enhance your day-to-day computing experience, whether you're working or simply browsing the web. Most modern displays use Liquid Crystal Display (LCD) technology, which employs an LED backlight; in the past, Cold Cathode Fluorescent Lamps (CCFL) were commonly used for backlighting, but LED technology has largely replaced them. This section will guide you through the essential aspects of displays, including LCD and OLED panels and their recommended uses, resolution & aspect ratio, refresh rate, brightness, and the various features and connectivity options available. Next, we will cover keyboard and mouse options. By the end, you'll be equipped to select a monitor, mouse and keyboard for your Non Gaming PC build.

Panel Types

Twisted Nematic (TN) LCD Panels: Having been around since the 1970s, TN panels were among the first types of LCD technology to be used in computer monitors. They became popular in the 1990s and 2000s due to their affordability and fast response times. TN panels have poor off-angle viewing, resulting in color and contrast shifts when viewed from different angles, and they have lower color accuracy compared to other panel types. The latest TN panels now offer improved color accuracy and better viewing angles than earlier models, making them more competitive with IPS and VA based monitors. For the Non Gaming PC on a budget, TN panels offer fast response times at a lower cost compared to other monitor types.

Vertical Alignment (VA) LCD Panels: VA panels offer improved color accuracy and better contrast ratios compared to TN panels. They became popular in the 1990s for general use and multimedia applications due to their good overall performance, though they had slower response times than TN panels and some ghosting issues. They excel in general use with a focus on contrast and HDR performance, providing deeper blacks than TN and IPS monitors. VA panels are also well-suited for use in curved displays where an immersive experience is the goal. Technology keeps improving, and VA panels continue to excel in contrast and black uniformity, while they have also seen improvements in response times and color accuracy. When used in a darkened room for watching movies, a VA panel can deliver a satisfying viewing experience.

In-Plane Switching (IPS) LCD Panels: IPS panels were developed to address the limitations of TN panels, particularly in terms of color accuracy and viewing angles. They quickly became the preferred choice for professional applications such as graphic design and photography due to their superior color reproduction and consistent image quality. IPS panels are best for productivity and color-critical work due to their excellent color accuracy and good off-angle viewing. With developments in technology, IPS panels have improved in response time and refresh rate. With the advancements in color accuracy and consistency, and faster response times, IPS panels are a solid choice for a broad range of productivity applications.

Organic Light-Emitting Diode (OLED) Panels: OLED marked a big leap forward in display technology. Unlike LCDs, OLEDs do not require a separate backlight, allowing for true blacks and high contrast ratios. They offer excellent color accuracy and viewing angles. OLED panels are best for mixed use, including media consumption and general computing. On the downside, they are more expensive than other panel types. They also have a potential for burn-in where static images can leave a permanent mark on the screen. OLED panels have a shorter lifespan compared to traditional LCD panels due to chemical degradation over time. OLED continues to develop better image cleaning techniques to combat burn-in. While still relatively expensive, OLED panels offer performance benefits in terms of color accuracy and contrast that are appealing for those seeking the best display experience.

Native Resolution

The native resolution of a monitor is the specific resolution at which the display is designed to produce the clearest and sharpest image possible. At native resolution, each pixel on the screen corresponds directly to a pixel in the image being displayed, ensuring 1:1 pixel mapping. This alignment avoids any interpolation or scaling artifacts that can occur when running alternative resolutions, which can introduce blurriness and reduce image clarity. Text and user interface elements are generally clearer and easier to read at the native resolution, providing the best overall usability. To maximize the benefits of native resolution, take into account the size of the monitor. For instance, a 27-inch monitor is a good match for a 2K (1440p) resolution, while a 32-inch monitor is better for a 4K display. Always aim to run your monitor at its native resolution and minimize scaling to ensure the best visual experience. Keep your primary use case in mind when determining the appropriate size and resolution of the monitor you select.

Avoid scaling a 4K monitor down to 2K

I want to make sure you're aware of the differences when scaling 2K, 4K and 5K native resolution panels. A 4K panel cannot be cleanly scaled down to 2K without degrading image quality. A little bit of math is required to explain:

A 2K (or 1440p) screen area of 2560x1440 equals 3,686,400 total pixels.
A 4K (or 2160p) screen area of 3840x2160 equals 8,294,400 total pixels.
A 5K (or 2880p) screen area of 5120x2880 equals 14,745,600 total pixels.

To scale 4K down to 2K, we have to divide:
8,294,400 ÷ 3,686,400 = 2.25. 2.25 is not a whole number. The resolution does not divide evenly, and the screen does not scale proportionally, forcing the monitor to fabricate or "interpolate" pixels to compensate for the mismatch in size. This results in a loss of clarity and sharpness.

To scale 5K down to 2K, it works as follows:
14,745,600 ÷ 3,686,400 = 4. 4 is an even number. It divides evenly, and the screen scales down in exact proportion. No pixels are lost.

Bottom line: A 4K monitor needs to be run at 4K resolution with no scaling for the best picture quality. This is why I recommend a minimum size of 32 inches for 4K monitors. At smaller sizes like 27 inches, a high pixel density 4K panel makes text and icons uncomfortably small for most users, who will need to use scaling, which compromises image quality.

Screen Resolution

Screen resolution influences image clarity, detail, and the overall viewing experience. It refers to the number of pixels displayed on the screen, represented as width x height (e.g., 1920 x 1080). A higher resolution means more pixels, resulting in more detailed images.

Full HD, or 1080p, has a resolution of 1920 x 1080 pixels and is commonly used for most applications, including office work and multimedia. It provides a good balance between clarity and performance. Moving up the resolution scale, 2K Quad HD (1440p) at 2560 x 1440 pixels offers better clarity and detail, which is nice for professional work where higher detail is beneficial. At the high end, 4K resolution (2160p) at 3840 x 2160 pixels offers excellent detail and sharpness, suitable for professional photo and video editing. For even greater detail, 5K resolution at 5120 x 2880 pixels is often used in professional environments for design and editing tasks.

Pixel density, measured in pixels per inch (PPI), is another factor that affects image sharpness and text rendering. Higher PPI on larger screens ensures that images and text remain sharp and clear. Larger screens with higher resolutions and high pixel density can be viewed comfortably from greater distances.

Different use cases benefit from various resolutions. In productivity scenarios, higher resolutions offer more workspace. Creative professionals, such as photographers, video editors, and designers, benefit from higher resolutions for precise detail and color accuracy.

When considering hardware compatibility, ensure your graphics card can support the desired resolution at acceptable performance levels. Also confirm that your monitor and GPU have compatible ports, such as HDMI 2.0 or DisplayPort 1.4, to support higher resolutions and the fastest refresh rates.

There are trade-offs between resolution and refresh rate. Higher resolutions often come with lower refresh rates, so it's a good idea to balance these based on your primary use case.

Aspect ratio also plays a role in the viewing experience. Aspect ratio is the proportional relationship between the width and height of a display or image, expressed as a ratio (e.g., 21:9). The standard 16:9 aspect ratio is the most common for general use and multimedia. Ultrawide aspect ratios, such as 21:9 or 32:9, provide a wider field of view, beneficial for multitasking and professional applications.

Refresh Rate

The refresh rate of a monitor, measured in Hertz (Hz), indicates how many times per second the display refreshes the image on the screen. A standard 60Hz monitor, which refreshes the image 60 times per second, is perfectly adequate for most everyday tasks like web browsing, office work, and media consumption. However, high refresh rate monitors, such as those with 120Hz, 144Hz, or even 240Hz, offer smoother and more fluid visuals. Video editors and animators may find high refresh rate monitors useful for previewing fast-moving scenes with greater clarity and smoothness. Professionals who work with CAD software or 3D modeling can experience more fluid manipulation of complex objects and scenes. General desktop users who use applications with rapid screen updates will appreciate the improved visual experience and reduced eye strain that a higher refresh rate can provide. While high refresh rate monitors are often marketed to gamers, their advantages extend to other fields and use cases.

Brightness and Nits

When selecting a monitor, one of the specifications to consider is brightness, often measured in "nits." The term "nit" is a unit of luminance that quantifies how much light a display emits over a given area. A higher nit value indicates a brighter screen, which can benefit your overall viewing experience. For everyday tasks such as web browsing or document editing, a brightness level of around 250–300 nits is generally sufficient under normal indoor lighting. However, for activities like video editing or working in brightly lit environments, a monitor capable of higher brightness, 400 nits or more, can enhance visibility and color accuracy. Higher brightness levels are particularly beneficial for HDR (High Dynamic Range) content, where the contrast between dark and light areas gives a more dynamic display.

Windows Display Scaling

Windows scaling options allow you to adjust the size of text, icons, and other interface elements to ensure they remain readable and usable on high-resolution monitors. As screen resolution increases, the density of pixels (PPI) also rises, which can make on-screen elements appear smaller and harder to read. Windows' scaling feature compensates for this by enlarging these elements proportionally, ensuring a consistent and comfortable user experience across different resolutions, at the expense of image quality. This can be useful for 4K and higher resolution displays, where without proper scaling, screen elements would be too small to read. By customizing scaling settings, you can achieve an optimal balance between screen real estate and readability, making high-resolution monitors practical for everyday use.

Sometimes, scaling can disrupt the functionality of applications. Certain custom business applications require Windows resolution to be set at 100% to operate correctly. Adjusting the scaling for these applications can lead to issues such as missing text, improper window sizing, or hidden dialog boxes. Before enabling Windows scaling options, ensure that the applications you use are fully compatible to avoid these problems.

Additional Monitor Features and Connectivity

When selecting a PC monitor, additional features and connectivity options can be beneficial. One such feature is High Dynamic Range (HDR), which improves the contrast and color range of the display. This technology makes images appear more vibrant and lifelike, especially useful for watching movies or photo editing.

Another useful feature is the ability to rotate the monitor into portrait mode. This is particularly beneficial for coding, reading long documents, or browsing the web. A monitor that supports portrait mode allows for greater flexibility in how you use screen space.

VESA mounting is essential for those who prefer to mount their monitor on a wall or an adjustable arm. VESA compatibility ensures that your monitor can be easily attached to a wide range of stands and mounts, giving you more control over your workspace.

Ergonomic adjustments like tilt, swivel, and height adjustments help create a comfortable and customizable viewing experience. These features allow you to adjust the monitor's position to reduce strain on your neck and eyes, which promotes better posture during extended use.

The screen surface finish, matte or glossy, is another consideration. A matte screen reduces glare and reflections, making it easier to view the screen in bright environments. On the other hand, a glossy screen can offer more vibrant colors and deeper blacks but may reflect light sources, which can be distracting.

Consider connectivity options when choosing a monitor. Common ports include DisplayPort (DP), HDMI, and USB Type-A and Type-C. DisplayPort is often preferred for its ability to support high resolutions and refresh rates. HDMI is widely used for its versatility and compatibility with various devices. USB Type-C is becoming increasingly popular due to its ability to transmit data, video, and power through a single cable. USB Type-A ports on monitors can also be handy for connecting peripherals like keyboards and mice.

Input: Keyboard and Mouse

When choosing a keyboard, one of the primary decisions is whether to opt for a membrane or mechanical keyboard. Membrane keyboards are affordable and fairly quiet. They use a thin, flexible membrane to register keystrokes, which can result in a softer, some would say mushy, typing experience. This design has a less tactile feel than a mechanical keyboard and doesn't last as long, especially with heavy use. Prebuilt computers often come with a membrane keyboard.

By contrast, mechanical keyboards are renowned for their durability and tactile feedback. Each key is equipped with a mechanical switch that provides a distinct, satisfying "click" or "bump" when pressed, which can enhance typing accuracy and speed. Long lasting mechanical switches make these keyboards a great choice if you do a lot of typing. Mechanical keyboards come in various switch types, each offering varying levels of resistance, noise, and tactile feedback. While they tend to be more expensive and make more noise than membrane keyboards, the performance and durability of a mechanical keyboard can justify the price.

When it comes to selecting a mouse, you have two main options: optical and laser. Optical mice use an LED light to illuminate the surface beneath them, which is picked up by a sensor to detect movement. This technology is accurate on most non-reflective surfaces and is well-suited for everyday computing. Optical mice tend to perform best on mouse pads or surfaces with a bit of texture, which helps the sensor track movement more precisely.

On the other hand, laser mice use a laser to track movement, which allows them to achieve higher levels of sensitivity (DPI) and accuracy compared to optical mice. This makes laser mice better when precise control and fast response are required.

When choosing a monitor, keyboard, and mouse, think about what you want to interact with every day. For monitors, TN panels are budget-friendly with fast response times, VA panels offer better contrast, and IPS panels provide superior color accuracy. OLED panels deliver top image quality but at a higher cost. Membrane keyboards are affordable but may lack durability and feedback, while mechanical keyboards offer a better typing experience. Optical mice work well for everyday tasks on textured surfaces, while laser mice can provide higher precision.

The quality and reliability of your monitor, keyboard, and mouse contribute to your satisfaction. When selecting these items, spend some time considering what devices will be the best fit for your Non Gaming PC build.

Summary 2.8: Monitor, Keyboard & Mouse

Recommended Configuration

27–32" IPS panel: 2K: 27", 4K: 32" at native resolution. See the monitor in person to make sure it's right before buying.

Mechanical Keyboard: High quality and reliable. A high quality mouse from a reputable brand. Don't skimp, you will be interacting with this hardware on a daily basis. Try them out before buying if possible.

120Hz Refresh Rate: Gaming and most workflows will benefit from a 120Hz or higher refresh rate. 60Hz is a solid choice if motion graphics are not a priority.

Technologies Supported

Wireless Input Devices: Look for Radio Freqency (RF) or Bluetooth connectivity for the mouse and keyboard, with USB recharging to avoid buying batteries.

Panel Types and Use Cases

TN: Favored for fast motion graphics.
VA: Emphasis on deep blacks and high contrast media.
IPS: Solid all purpose choice.
OLED: Great for immersive media like movies.

Additional Features to Consider

Connectivity: DP, HDMI, USB-A, USB-C. Some office oriented monitors feature a USB hub.

Adjustments: Height, tilt, swivel, and portrait mode.

Warranty Period

A 1–3 year warranty is common for desktop monitors. Labor may not be covered for the duration of the warranty.

Build Guide

Contents

3.1 Overview & Tools

In this section you'll get ready for your Non Gaming PC build, cover the tools you'll need, and go over a parts checklist so you'll be sure to have everything on hand when you start.

Build Overview

To get started, you'll need to have all the parts and a few tools together in one place. You'll need access to a table or bench large enough to lay the PC case down on its side. There should be a grounded outlet nearby with a surge strip plugged in to power the PC during testing.

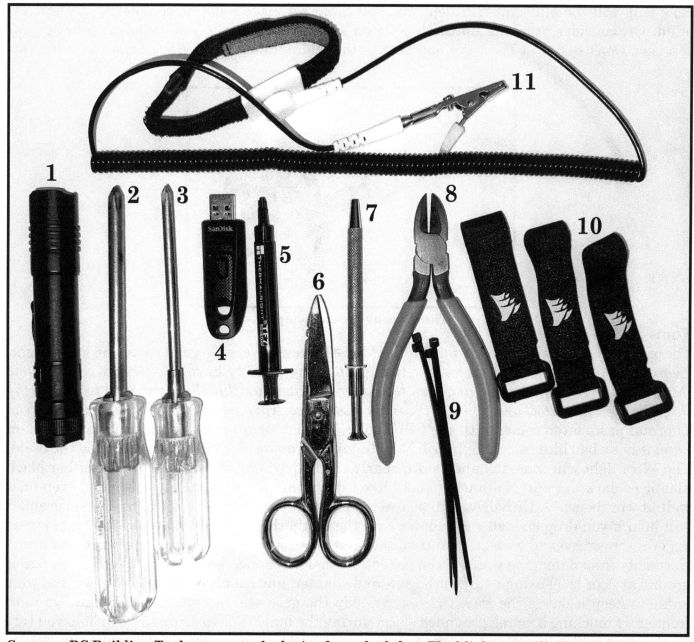

Common PC Building Tools, counter-clockwise from the left: 1. Flashlight, 2. #2 Phillips screwdriver, 3. #1 Phillips screwdriver, 4. USB drive, 5. Thermal paste, 6. Scissors, 7. Parts grabber, 8. Wire Cutters, 9. Zip ties, 10. Velcro ties, 11. Anti-Static Wrist Strap.

As you acquire the parts, open each box and check the contents to make sure that you received everything you're supposed to have. Take a little time to familiarize yourself with each part and any accessories. There may be an instruction sheet or pamphlet in the box with the parts. Don't discard the instructions, look them over and keep them, they may have useful information about installation or last minute changes made to the product. You'll want to save all the boxes at least until the purchase return window has expired. After that, it's still a good idea to keep the box for the CPU, GPU, case, and the motherboard, in case you want to sell these parts later or if you need to ship them back to the factory. To save room, store the smaller boxes inside the case box.

Try to get all your parts in hand and build the PC within the parts purchase return window. That way, if you have a dead on arrival (DOA) part, you'll be able to quickly exchange it for a replacement from where you bought it, without having to go through a manufacturer's return merchandise authorization (RMA) process, which can take a while and could delay your build.

How long will the build take? It depends. You could set aside a day to complete your first PC build. Or, consider breaking up the build into several stages that you complete over several sessions. Don't rush it. I know I've had to burn the midnight oil on more than one tricky build.

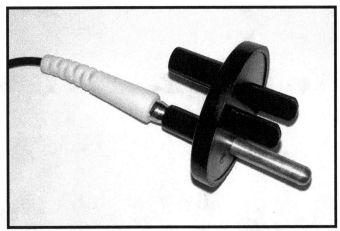

Three prong ground adapter.

Tools

The only tools you really need are the Phillips #2 & #1 screwdrivers, thermal paste and a flashlight. Everything else is nice to have, but optional. You may already have the screwdrivers. *Screw threading tip: Insert the screw and turn it counter clockwise until it clicks one time to line up the threads, then turn clockwise to snug the screw into place. This avoids cross threading the screw.* Thermal paste often comes with the CPU cooler, but having an extra tube of thermal paste on hand is not a bad idea. A flashlight will help illuminate the inside of the PC case during the build. The extra light will come in handy since nearly all the parts inside the case will be either black plastic or dark in color. Not much light will get down into the case during assembly, even on a well lit workbench. A USB drive will be needed if you have to update the BIOS. A parts grabber will help if you drop a small screw down onto the motherboard, or you can use the magnetized tip of a screwdriver to assist. An anti-static wrist strap and static mat can help prevent static electricity from damaging your PC components. Consider connecting the wrist strap to an outlet ground at floor level using a three prong ground adapter, and then looping the strap around your ankle to keep it out of the way. Or, you can skip the grounding strap, but make sure you are frequently touching a metal, grounded object during the build. Velcro or zip ties can help you tidy up the loose cables to neaten the appearance of your build. If you go for zip ties make sure you've got snips or scissors to cut them when needed. I prefer Velcro because its reusable and adjustable without having to cut anything inside the case.

3.2 Case & Fans

First, remove all the outer panels from the case. Case side panels are usually secured with captive thumb screws. After loosening the screws, slide the panel back and away from the case body to remove the panel. The thumb screws will remain attached to the panel. Lift the panels away and put them aside. You can temporarily store them in the case shipping box for safe keeping. Front panels and air filters are often friction fit. Remove and store these as well.

When mounting fans, the orientation is important. Fans have an open intake side and a framed exhaust side. Some PC fans have arrows marked on the frame to indicate the direction of airflow and blade rotation. Install the fans to support positive pressure airflow in the case. Configure more fans for intake of fresh air than for exhaust. Slightly positive air pressure inside the case prevents pulling in dust and debris through the crevices.

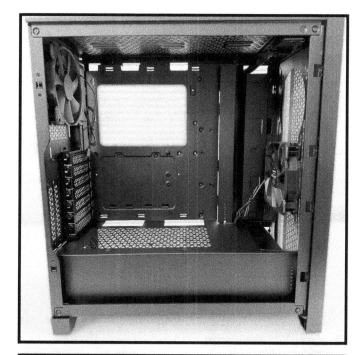

Left: Exhaust side, frame circled. Right: Intake side.

Your fans will come with appropriate fan mounting screws. When attaching the fans to the case, the screw goes through the slot in the case and into the mounting holes in the corners of the fan body.

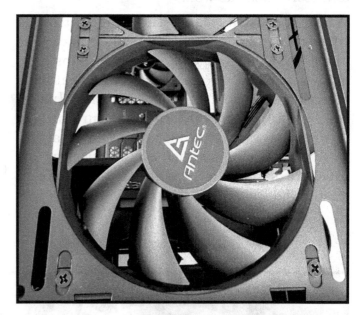

The case I'm using in this build, the Corsair 4000D Airflow, came with 2 fans of the 3 pin connector variety, which are OK but I wanted a little more control so I purchased a 5 pack of Antec PWM fans. I replaced the 3 pin fans with PWM units and added 3 more PWM fans for intake at the front of the case.

Wire leads are attached to each fan, which will connect either to a splitter or directly to the motherboard fan headers. Orient fans during installation so that the cables are closest to the outside edge of the case or motherboard mounting tray. Avoid crossing the leads over the fan or other components. This will result in the shortest wire run and a neater build.

Install fans so that this corner is closest to the motherboard tray for the shortest cable run.

Now is a good time to install any 2.5" SSD's or 3.5" hard disk drives in the case.

Note: You will leave these drives disconnected until after Windows is installed on your primary NVME drive.

Most cases come with a drive bay area or trays to accommodate these drives. The SSD tray shown here has a thumbscrew for easy access.

The Corsiar 4000D Airflow case has an HDD bay with removable "sleds" that can hold 3.5" drives without using screws. The plastic sled is easily removable from the bay. The HDD drive bay is also removable by loosening the thumbscrews.

Line up the holes in the side of the drive with the mounting pins on the sled rails and insert the drive between the sled rails. Once in place, slide the sled back into the bay where it will click into place. 2.5" drives are attached to the case with the mechanism on the back side of the motherboard tray. The case is now finished and you can now set it aside.

3.3 CPU Installation

The first component you'll install on the motherboard is the CPU. You'll install the CPU on the motherboard outside the case, with the motherboard resting on the box that it came in. This allows more room to work and see what we're doing during the process. It also helps protect against damage from electrostatic discharge.

Before starting, here is a brief overview of the process to install an LGA socket processor, which is basically the same for an AMD or an Intel CPU: 1. Lift the lever. 2. Lift the metal socket cover. 3. Align the CPU to the correct corner and place it into the socket. 4. Close the socket cover over the CPU. 5. Close and secure the socket lever. If the plastic socket cover is still in place, it will dislodge now as the lever is closed. Retain the plastic cover. Store it in the CPU box. Should you later need to remove the CPU for any reason, the cover will be required to protect the socket.

Open the CPU outer box and leave the CPU inside its plastic capsule for now.

Put the motherboard on top of its box on your work surface. This provides a convenient position to work on the motherboard.

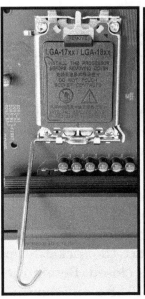

First, release the socket lever. Push the lever down and away from the catch and lift it up. The lever is now in the raised position.

Raise the socket cover up from the tab on the upper left to expose the CPU socket. Note the triangle in the corner of the CPU socket.

The CPU socket is open. The CPU orientation corner is circled at the right in this photo.

Remove the CPU from the plastic capsule. Align the CPU with the socket using the triangle marked on the corner of the CPU and the notches in the CPU board and socket. Gently lay the CPU into the socket. No force.

Lay the socket cover back down over the CPU.

Push the lever back down and latch it into place under the catch where it was before. As you do this, the plastic CPU socket cover will pop off if it hasn't already. As noted above, save the black plastic socket cover for use if you have to remove the CPU from the motherboard.

The CPU is now installed. Great job!

3.4 RAM Installation

When installing memory sticks, first figure out what slots are to be populated first. It may be indicated right on the motherboard near the slots. If not there, the motherboard manual will tell you exactly which slots need to get populated first.

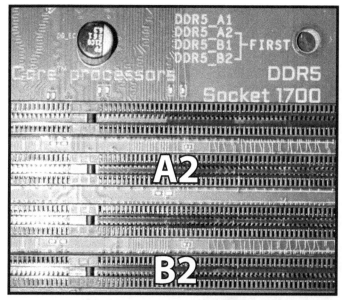

In this example, the slots A2 and B2 are the first slots to receive memory modules. This puts them at the end of the memory channel, which prevents noise and signal echoing that may occur when the 1st slot is populated and the second slot is empty.

Open the latches at both ends of the slots.

Carefully line up the notch in the memory module with the corresponding notch in the slot and push the module straight down from the top. It will require some pressure.

As the module seats, the latches at the end will close and you'll hear them click shut.

Once you have one module seated, the other will go in the same orientation.

After both modules are seated, confirm that the latches are fully closed. The RAM installation is now complete.

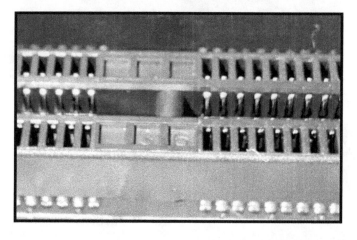

3.5 M.2 NVMe SSD Installation

NVME SSD's are installed on the motherboard in slots designated for that purpose. Most motherboards have several M.2 slots, one next to the PCIe 16x slot intended for the GPU, and 2 or 3 others not too far away. Sometimes one of these slots is shared with the SATA ports on the motherboard, consult your motherboard manual to see which one, if any, are shared. If you use a shared slot for an NVME, this means that one or more of your SATA ports will be inactivated while an NVME is in the shared slot.

The **M.2 slot closest to the CPU** is often on the same bus as the CPU and may be a bit faster than slots positioned farther away which are controlled by the chipset bus. Use this M.2 location for your primary NVME drive. This will become your **C:** drive after Windows is installed. *If you have multiple M.2 NVME drives, wait to install the others until after Windows is set up on your primary drive.*

If there is an M.2 heatsink on the motherboard, remove the screw securing the heatsink with the appropriate size Phillips head screwdriver.

Remove the film covering the adhesive on the heatsink. Also remove the film covering the adhesive on the motherboard at the NVME slot. Check the boss where the NVME drive will mount. It will either be a screw or a spring loaded latch. If there is a screw there remove it now.

The NVME drive has a notch in the connector end. Align the notch with the notch in the slot. Angle the NVME drive forward into the connector until it clicks into place. It will now be suspended in the slot.

Push the NVME drive down onto the mounting boss and secure the latching mechanism or install the screw to hold the drive in place.

Place the heatsink on the NVMe. Make sure the film covering the adhesive has been removed. The heatsink may have a tab at one end that inserts into a slot, and a screw at the other end to hold it down.

Primary M.2 NVMe installation is complete.

After Windows is installed, you can install other M.2 drives and connect any 2.5" SSDs, or HDDs if you have them. You will probably have to remove the GPU to install those other M.2 drives later. This is a small inconvenience but it makes your Windows installation easier.

GPU clearance note: In this example, the NVMe heatsink is positioned next to the PCIe 16x slot. Not very often, but in some instances, the height of the heatsink interferes with the installation of the GPU. In the example shown, using the heatsink provided with the motherboard and an NIVDIA GPU, there was plenty of room. If you think you may have a clearance issue due to excessive height of your NVME heatsink, you can test fit the GPU with the heatsink in place.

To test fit the GPU, remove the edge cover from the card and open the retaining latch at the end of the PCIe 16x slot. Fit the GPU down into the slot and verify that the GPU seats fully in the slot and that the retaining latch closes fully. Observe the space between the bottom edge of the GPU and the top of the NVMe heatsink. If the heatsink touches the GPU or interferes, you'll need to replace the heatsink with a low profile unit. Be sure to open the retaining latch at the end of the PCIe slot before removing or inserting the GPU.

3.6 Thermal Paste & CPU Cooler

In this part of the build, you will spread the thermal paste onto the CPU and install the air cooling tower. The thermal paste is what makes the connection between the metal body of the cooling tower and the integrated heat spreader (IHS) of the CPU. Without thermal paste and a cooler, the CPU will quickly overheat and be ruined.

I prefer to spread the paste over the entire surface of the IHS as opposed to using a bead or line of paste. I used a semi-flexible piece of plastic to spread the paste, but you can use your finger as well. I added a bead of paste along one edge of the CPU and spread it across the IHS.

The first pass wasn't quite enough, so I added another bead of paste at the other end and spread it out until I was satisfied that the entire CPU was covered with a layer of thermal paste. Try not to be stingy with the thermal paste, but don't make a mess of it either. Slightly too much is better than not enough.

After spreading the paste, install the bracket for your cooler onto the motherboard. In the featured build I'm using a Thermalright Peerless Assassin air cooler kit supplied with 2 fans and an LGA1700 socket bracket designed for Intel CPU sockets. Similar kits are available for AMD processors. I installed the bracket from the back. The bracket has an adhesive pad that sticks the bracket to the back of the motherboard during installation.

I added standoffs from the kit for each of the corner posts and installed the brackets on top of the posts with the screws in the kit. Note the orientation of the posts on the brackets. They are aligned to mount the cooler and fans facing front to back in the case.

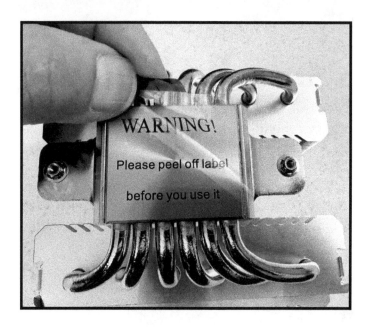

Remove the plastic film covering the mating surface of the air cooler tower and mount it onto the brackets and down to the CPU.

Tighten the screws on the cooler from side to side, keeping the pressure even until both sides are snug, no need to over tighten.

Mount the fans on the cooling tower fins so the frame side of the fan is against the fins. You want to push air into the fins and have that air moving from front to back in the case. Orient each fan so that the connector lead is closest to the CPU fan headers on the motherboard and the framed side of the fan is touching the fins. Attach the fans to the fins using the spring clips supplied with the kit. Hook the springs through the holes in the fan body and snap them over the fins. One fan may have to ride a little higher than the other to avoid contacting the memory modules. Adjust the fit of the fans so they don't contact other motherboard or case components.

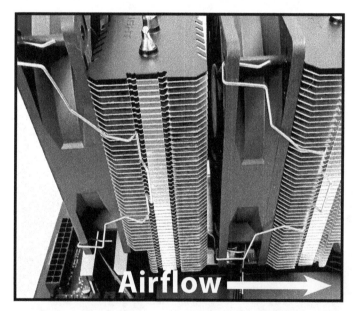

Once the fans are in place, attach the 2 leads to the splitter supplied with the kit, and connect the lead to the CPU_FAN header on the motherboard to control both CPU cooler fans as one unit. The connector is keyed to only fit one way.

The CPU cooler installation is complete.

3.7 Pre-Assembly System Check

The Pre-Assembly System Check is the first test of your computer build and lets you know that your build is functioning and ready to install into the PC case. It also confirms that the motherboard passes its own Power-On Self-Test (POST Test). The advantage of doing this now is that it's much easier to reseat and adjust components while the motherboard is outside the case. At this point, the motherboard has the CPU, thermal paste and CPU cooler, RAM and NVMe storage installed. To run the test for this build outside the case, you'll add the power supply connections and the GPU. If you're adding a PC speaker to your build, add that now as well. Go ahead and put on your ESD anti-static strap now, or make sure you are grounded before continuing.

With the motherboard positioned on top of its box on your work surface, get out your power supply and monitor, and set them on the bench nearby.

If the CPU cooler fans aren't connected to the motherboard yet, connect them now to the CPU_FAN header.

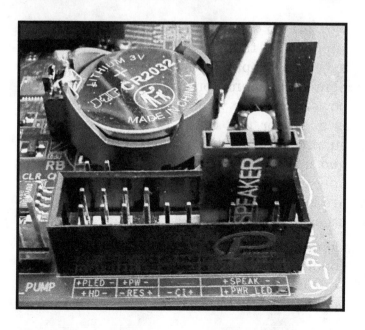

If you're adding a PC speaker, locate the +SPEAK- header on the motherboard. It's probably grouped together with the power and reset switch pins. The PC speaker has a 4 pin connector but only 2 are connected. Simply push it down onto the pins.

You can plug in a USB keyboard and mouse at the motherboard rear I/O panel now. A wireless RF mouse and keyboard will work but the mouse may be less responsive in the BIOS than a wired unit at this point.

Prepare the GPU for temporary installation by unwrapping and removing any protective film, tape, or covers. Remove the PCIe edge connector cover and any needed DP/HDMI socket plugs. The photo highlights the PCIe retention notch on the GPU edge connector, which engages the latch in the motherboard's PCIe slot. Always open or release this latch when installing or removing the GPU. Refer to the photo of the latch at the end of section 3.5.

Position the GPU over the PCIe 16x slot, make sure the latch on the slot is pushed back and press the card downward into the slot. It will click into place. Confirm that it's fully seated and there are no obstructions from heatsinks or other components. Let the metal tabs of the GPU hang off the side of the box supporting the motherboard.

Now attach the power cables, first to the power supply and then to the motherboard. You'll need the 24 pin motherboard connector, the 8pin CPU connector(s), and the appropriate power cable for your GPU. If the GPU cable is a 4+4 or 6+2 connector from your power supply, position the connectors side by side and insert them into the connector. They will only go one way and you'll notice some edges are keyed such that some are squared off and others are rounded.

Connect your display to wall power and connect the appropriate video cable between your monitor and one of the ports on the rear edge of the GPU.

After everything is plugged in, confirm that the power supply switch is in the OFF position, and connect the PSU to wall power. Your setup should look something like the photo at right.

Check all the connectors one more time and move the PSU power switch to ON. Nothing happens.

The motherboard won't power on automatically outside the case when you turn on the power supply. You'll need to manually power it up by pressing the motherboard's power switch, if available, or by momentarily bridging the power switch header pins (+PW-) with a flat blade screwdriver.

If your motherboard has an onboard power switch, go ahead and turn it on. Otherwise, locate the +PW- header on your motherboard and have a flat blade screwdriver ready. Touch the + and - pins of the power switch header with the screwdriver for a moment. This brief connection will signal the motherboard to turn on. There is no line voltage there so it's perfectly safe.

The motherboard should power up now, with LED lights illuminating and the CPU fan spinning. If you hear a single beep from the PC speaker, it means the board has passed the POST test. Many motherboards now use LED lights to indicate the status of CPU, RAM, and other functions, check your manual for details.

At this point, your monitor should display the BIOS. Verify that your memory, M.2 NVME drive, and GPU are recognized, and check the CPU temperature, it should be between 35-45°C or lower and stable. Avoid changing any BIOS settings except for the date and time.

If everything appears to be working correctly, you're ready to install the motherboard into the case. Power down and remove the GPU (don't forget the latch). **Leave the CPU power connected to the motherboard. Disconnect the cable from the power supply and drape it over the top of your motherboard assembly. Do this, because the CPU power connector will be difficult to reach once the motherboard is placed into the case.** You could also leave the 24 pin motherboard power connector attached as well, but it's pretty easy to reach.

Troubleshooting the Pre-Assembly System Check

CPU Temperature Issues:

In the unlikely event that the CPU temperature is rapidly rising now, especially if it's over 45C and rising, immediately switch off the power at the PSU. There may be a problem with your CPU cooler installation, either with the thermal paste or with the CPU cooler hardware setup. Disconnect the CPU fans from the motherboard and check the CPU cooler hardware setup. Remove the CPU cooler and reapply the thermal paste, reinstall the cooler, then try the POST test again.

Video Issues

Another issue may be video: The display is blank, no video showing, but the CPU fans are on.

Is the monitor turned on? Is the display source set correctly to the input you're using? If your cable is plugged into DP1, make sure the monitor source is set to DP1.

Reseat the video cable at both ends then check again.

If still no video

> Turn off the power supply
>
> Disconnect the video cable from the GPU
>
> Disconnect the GPU power cable at both ends.
>
> Reseat the GPU card in the PCIe x16 slot. Remember the little latch on the slot.
>
> Reconnect the GPU power cable at both ends and make sure it's fully seated.
>
> Reconnect the video cable.

Restart the test.

If still nothing, plug the video cable into video port on the motherboard I/O panel instead of the GPU. Is there video now? This should begin to narrow down where the problem may be.

Below are some more things to check if the system is not powering up

Ensure the power supply (PSU) is switched on.

Confirm that the 24-pin ATX motherboard power cable and the 8-pin CPU power cable are secure and fully seated.

Verify that the power cables for the GPU are properly connected.

Check that the CPU fan is connected to the CPU_FAN header on the motherboard. Some boards will not start if there is no fan connected for the CPU.

Ensure that the RAM sticks are fully seated in the correct slots according to the motherboard manual. If using multiple sticks, try booting with only one stick installed and then test the others individually.

Make sure no cables are obstructing fans or components.

Listen for any beep codes from the motherboard speaker, if installed, or check the motherboard's diagnostic LED display for error codes. Refer to the motherboard manual for decoding these signals.

Reset the motherboard's BIOS settings by using the CMOS reset button. Consider updating the BIOS to the latest version. See Appendix A: BIOS Update for more information.

3.8 Mounting the Motherboard

Now it's time to install the populated motherboard into the case. If you haven't prepared the case by installing the fans and any SSDs or hard disk drives, go ahead and do that portion of the assembly now because there just isn't much room to work inside the case once the motherboard and CPU cooler are taking up most of the space. Make sure that the fan cables are routed out of the way and are not laying on the motherboard tray or hanging across or into the rectangular I/O panel space at the back of the case. Arrange the CPU power cable such that it won't interfere with lowering the motherboard assembly into the case.

I/O Shield Note: The motherboard shown in this build has its I/O shield attached to the board. Your board may have a separate I/O shield that covers the ports on the back of the motherboard. This type of I/O shield fits into the rectangular opening in the back of the case. Make sure to fit the I/O shield in place on the case if your motherboard has this arrangement.

Position the case on your bench backside down, with the motherboard tray facing up and case cables tucked out of the way, they will be routed later. On the motherboard tray you'll notice some posts for mounting screws and one locating post without a screw hole. These will all line up with the holes in the motherboard. Your case may even be marked with a legend to indicate what size (ATX,ITX, etc...) motherboard the mounting locations are used for. The center locating post is marked by an arrow in the photo.

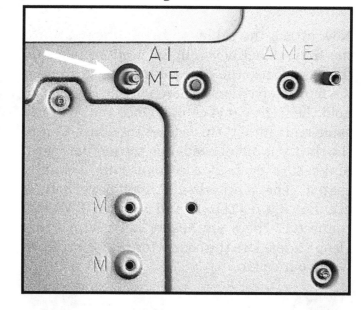

Now it's time to lift up and lower the motherboard into the case. You can carefully use the CPU air cooling tower as a handle if you like. The center hole of the motherboard will go over the locating post in the center of the tray. Carefully lower the board down and watch for the post through the center hole. It helps to have plenty of light for this step. A flashlight can be handy right now to confirm that all the mounting holes are positioned over the posts after the board is set in place.

Once the motherboard is on the center post, you can begin placing the flat washers and threading the hold down screws into their posts. At each post, place a flat washer over the mounting hole and put the screw into position. Turn the screw counter clockwise until you hear it click one time so you know it's lined up in the threads, then loosely thread in the screw at first, then move on to the next one, until all of the screws are started, then go back snug them all down. This will leave you a little slack to allow all the mounting holes to line up. A magnetic tip screwdriver will help in the hard to reach areas.

Now attach the cables to the motherboard for the front panel ports, lights, and switches on the case. Raise the case to standing position if it's more convenient. There may be a socket to hold the front panel cables. Once you've fit the cable ends into it, this socket then inserts into the the front panel header on the motherboard. Make sure each of the connectors are fully seated. The motherboard may have both a regular 2 pin PLED+/- and a 3 pin PWR_LED connector. The 3 pin header is for supporting sleep states. Use the connector that is supplied with your case.

Attach all the fan connectors to the motherboard. For this build I grouped the 3 front intake fans on a 3 way connector and attached them to the SYS_FAN1 motherboard header. The CPU fans are grouped together and connected to the CPU_FAN header. Feed any excess cable through the case and away from the fans. The exhaust fans at the rear and top of the case are grouped together and connected to the SYS_FAN2 motherboard header. Make sure now that none of the fan leads interfere with the operation of the fan blades.

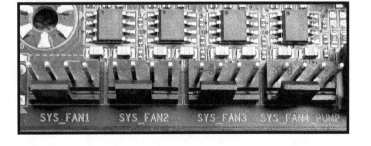

3.9 Power Supply, GPU, and Cables

For this build I selected a Corsair RMx Shift Series RM1000x power supply. These fully modular units have the power connectors on the side of the power supply instead of on the front, reducing the cable clutter and improving airflow inside the case. I paired this PSU with a Corsair case to ensure that it would fit. A traditional PSU will have cables in the front, allowing for shorter cable runs to the various connectors and wide compatibility with a variety of PC cases.

NOTE:
ONLY USE THE CABLES THAT CAME WITH YOUR POWER SUPPLY.
DO NOT MIX CABLES BETWEEN POWER SUPPLIES, EVEN THE SAME MODEL.
See the *Parts Guide: Power Supply* section for more information.

If you're using a traditional PSU with front-facing cable connections, attach the necessary cables before mounting it in the case. Position the PSU with the fan facing down to align with the bottom case vent. If your case doesn't have a bottom vent, install the PSU fan-side up to act as an additional exhaust. Secure the PSU to the case with the provided screws.

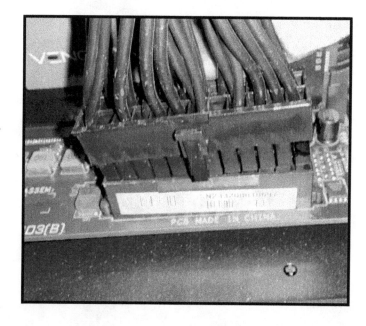

Now begin connecting the power and data cables inside the case, starting with the 24 pin power cable to the motherboard. The 8 pin CPU power cable should still be connected to the motherboard as outlined near the end of the POST test section.

Get plenty of slack cable inside the case to make positioning the cables easier.

When working around the CPU air cooler fins, drape a piece of cloth or paper over the fins to allow easier access.

Connect the SATA and Power cables for the HDD or SSD to the motherboard and PSU, but leave them disconnected from the drive until after Windows is installed. Using the straight end of the cable, I connected the HDD to the 4th SATA port, SATA3_4, avoiding any shared M.2/SATA ports. Refer to your motherboard manual to confirm which SATA port you need.

You can use the right angle connector end of the cable for the hard disk drive SATA connection when you are ready to connect it after Windows is installed. Just leave the data cable and the power cable shown below loose for now. You'll connect these after Windows has been installed.

The power connector for the HDD connects to a SATA/PATA port on the power supply and mates to the HDD with another right angle connector.

If you haven't already, go ahead and hook up the USB-3, USB Type-C, and HD Audio cables from the case to the motherboard. Refer to your motherboard manual to identify the location for these connectors on your board.

These connectors are all keyed so that they only go in one way. This is the USB Type-C.

And the HD Audio cable.

To install the GPU in the case, first remove the expansion slot cover(s). If your GPU is wide, you'll need to remove 2 of the slot covers near the PCIe x16 slot. This case has thumb screws, but it's easier to reach with a screwdriver.

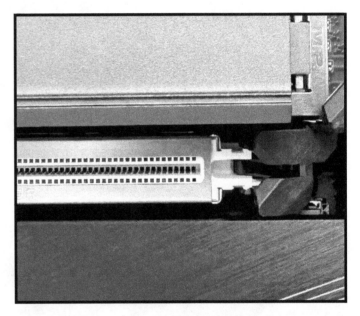

Confirm that all the plastic wrap and tape is removed from the GPU. Rock back the small latch at the end of the PCIe x16 slot. Insert the card into the PCIe x16 slot. Push it firmly down and the latch will close. Use the screws removed from the slot covers to attach the GPU to the case.

Use the screws removed from the slot covers to attach the GPU to the case.

Connect the GPU power cable now, taking care to fully seat and latch it at the GPU. If the connector is a 6+2 or 4+4 connector, look at the connector and align it so it will fit. They are keyed so they will only fit one way.

Manage the cables on the back side of the case behind the motherboard tray so that you can close easily close the case panel without binding any of the cables. Your case will probably have tiedowns and raceways for cables.

Attach the side panels to the case with the thumbscrews. Make sure that none of the cables are interfering with any of the fans.

3.10 First Power On

Now everything is ready for the first power up of the new build in the case. Follow the steps below to make sure everything is running correctly.

Set the computer case in the place where you want to run it. Confirm that there is space for ventilation. Make sure the PSU fan at the bottom of the case is not blocked by carpet.

Setup your display and connect it to power. Connect the display data cable from the monitor to the back of the GPU. Most likely this is a DP to DP cable. Confirm that you are connecting the display to the GPU and not to the DP port on the motherboard I/O panel.

Connect your mouse and keyboard. If you're using a wireless mouse and keyboard, set it to RF unless you know your BIOS supports bluetooth.

Connect the power supply to wall power. Flip the power switch on the PSU from the off (0) to the on (1) position. The PC is now ready to power up.

This is the moment of truth, press the power button on the case. The system should turn on, and there will probably be some lights on the motherboard. If you installed a PC Speaker, you'll hear a short chirp. Then the BIOS screen will appear on the display without intervention since the BIOS will find no bootable devices.

On the BIOS screen, confirm that all your components are present and recognized. The CPU temperature should be close to the temperature you saw during the Pre-Assembly System Check.

Make sure all the case fans and CPU cooler fans are running. Since there is no real graphics load on the GPU, its fans will probably not be running and this is OK.

Don't make any changes to the BIOS, such as enabling XMP, until after Windows is installed. If you haven't set the date and time in the BIOS yet, go ahead and do that now. Your BIOS may look similar to the image below.

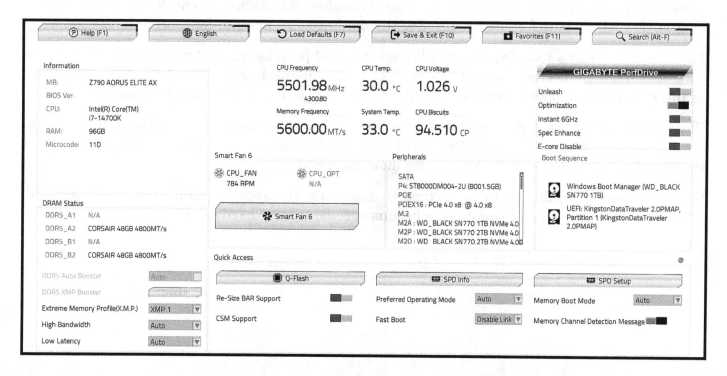

Troubleshooting Checklist

Following is a list of items to check and things to do if your freshly built PC isn't working properly. If your build passed the Pre-Assembly System Check, the hardware should be fine now, just a matter of checking the cables, connections, and fitment of the various components in your build.

Ensure the power supply (PSU) is switched on.

Confirm that the 24-pin ATX motherboard power cable and the 8-pin CPU power cable are securely connected. Verify that the power cable for the GPU is fully seated on the card.

Check that the monitor is powered on and connected to the correct output on the GPU or motherboard. Is the display source set correctly to the input you're using? If your cable is plugged into DP1 on the monitor, make sure that the monitor source is set to DP1.

If the PC won't switch on, check the front panel connectors, especially the power switch connector, to ensure it is properly seated on the motherboard.

Check that the CPU fan is connected to the CPU_FAN header on the motherboard. Some boards will not start if there is no fan connected for the CPU.

Ensure that the RAM sticks are fully seated in the correct slots according to the motherboard manual. If using multiple sticks, try booting with only one stick installed and then test the others individually.

Ensure the GPU is fully seated in the PCIe slot and the latch is closed.

Make sure no cables are obstructing fans.

Ensure the motherboard is correctly installed with all standoffs in place and that no loose screws or metal objects are causing a short circuit under the board.

If your motherboard requires an I/O shield, verify that it is aligned with the case and not shorting the motherboard or pinching any cables.

Listen for any beep codes from the motherboard speaker, if installed, or check the motherboard's diagnostic LED display for error codes. See the motherboard manual for more information.

Try a back to vanilla test: Disconnect all non-essential peripherals and drives, leaving only the CPU and RAM connected, connect the monitor to the motherboards video port, and try to boot again. If the system boots, reconnect components one at a time to identify the problematic part.

Clear the CMOS: Reset the motherboard's BIOS settings by using the CMOS jumper or use the BIOS reset button if your motherboard has one. Consider updating the BIOS to the latest version. See Appendix A: BIOS Update for more information.

If all else fails, consult forums or customer support for the specific components in your build. They may have additional insights or solutions specific to your hardware.

Appendix A: BIOS Update

Overview of the BIOS Update System

Motherboard manufacturers update the BIOS for their products and make these updates available for download on their websites. Customers download the BIOS files and, following the manufacturer's directions, install the new file onto the motherboard by "flashing" it. These updates correct problems in previous BIOS versions and update existing motherboard features, such as enabling support for a new CPU model.

Do I Have to Update the BIOS?

No, BIOS updates are not required, and you may never need to update it. Unless you've identified a specific problem or are upgrading the RAM or CPU, the BIOS that shipped with the board from the factory will usually be sufficient to support the hardware available when the board was released. If you have a recently released CPU or new RAM technology, check the manufacturer's website to see if your board supports the new hardware. While you're there, see if they have any BIOS updates that apply to specific hardware or software you will be using and consider getting the update if it impacts you.

General Process to Update a BIOS

Things You Will Need:

- USB Drive formatted to FAT32

- Motherboard and the manual (download if you don't already have it)

- Power Supply and cables

- Anti-static strap

- Access to a computer

1. Locate, download, and extract the BIOS file to a folder on a computer.

2. Copy the BIOS file(s) to a USB flash drive according to the manufacturer's instructions.

3. Insert the flash drive into the correct port on the motherboard and perform the flash process, following the manufacturer's instructions. Do not allow the power to go out during the flash update. If the power goes out during the update, it's possible that the BIOS will be corrupted and you won't be able to roll it back to the previous version. If this happens, contact the manufacturer.

4. Allow the update process to complete (should be less than 10 minutes), then remove the USB drive and restart the system.

Step-by-Step Process to Update the BIOS on a Gigabyte Motherboard Using Q-FLASH

1. Identify the motherboard by product name and version on the box and on the board. In this example, the board is a Gigabyte Z790 AORUS Elite AX (rev. 1.x), as identified on the box and on the motherboard itself.

2. Locate, download, and extract the BIOS update .zip file to a computer.

3. Find the file named after the BIOS update (e.g., examplefilename.FH), rename it to GIGABYTE. BIN, and then copy that file to the root folder of a USB drive formatted FAT32 as per the instructions.

4. Ground yourself or attach your anti-static strap to your arm or ankle. Take the motherboard out of its box and place it on top of the box.

5. Insert the USB drive into the designated port on the board. This Gigabyte board has a port specifically designated as the Q-FLASH port.

6. Set up your power supply, connect the 24-pin and 8-pin power cables to the motherboard.

7. Plug in the power supply to wall power. Turn on the power supply.

8. Press and hold the Q-FLASH button on the motherboard for one second.

9. An LED light at the Q-FLASH switch will begin flashing.

10. Various lights on the motherboard will flash periodically for about 7 minutes, then the lights will go out.

11. After about 10 minutes, turn off the power supply, disconnect the cables from the motherboard, and unplug the power supply from the wall.

Now the BIOS is updated. Enter the BIOS screen at startup and double check that the update has successfully installed the new version.

After updating the BIOS, if you experience any issues or instability, you may consider clearing the CMOS to reset the settings to their defaults for the new version. Check the manual for your motherboard for the clear CMOS procedure.

Windows Guide

Contents

4 Windows Guide

Installing and configuring the operating system is a relatively straightforward process. In the following section, I'll provide guidance on getting Windows installed and configured on your Non Gaming PC. Rather than walking you through every installation screen, since most are self-explanatory and easy to follow, I'll discuss a few steps that will help you understand the process.

4.1 Windows Installation Media

There are several methods for getting the Windows installation files. If you have access to a Windows computer, you can go to Microsoft's website (*https://www.microsoft.com/en-us/software-download/*) and download the media creation tool. After downloading the media creation tool, you can create the Windows installation media on a USB drive. You'll need a USB drive that is at least 8GB in size. The USB drive will be erased and formatted during the creation process, so make a copy of any data you need from that USB.

If you have access to an Apple computer, a good way to create Windows installation media on macOS is by using the Media Creation Tool within a Windows virtual machine running in Parallels. Parallels is a virtual environment that allows you to run Windows on a Mac. Setting it up is straightforward, and you may be able to run it as a free trial.

Another way to get Windows installation media is simply to purchase it from a retail outlet. This is especially useful if you do not have access to a computer. You'll receive a bootable USB drive containing the Windows installation files along with a registration key that will allow you to register your copy of the Windows operating system.

You can install Windows and run it for an unlimited amount of time, at no cost, without a registration key. Until your copy of Windows is registered, you will see some notices about activation and some of the personalization features will be limited. But the core functionality of the operating system will not be affected. You'll be able to install software and do just about everything that a registered Windows installation can do.

Caution: Make sure you get your Windows operating system files and registration keys directly from the official Microsoft website or from an authorized retailer. This ensures that you'll receive genuine software and you'll avoid potential security risks associated with unverified sources.

4.2 Installing Windows

You will be setting up Windows on your primary M.2 NVMe drive, the one installed closest to the CPU. At this point, that should be the only drive installed in the system. Connect the Windows media USB to your PC and reboot. The BIOS will find the bootable USB and Windows setup will begin. After setting the language and currency, setup will ask you to activate Windows with a product key. If you have a product key, go ahead and enter it now. If you don't have a product key, just click "I don't have a product key" to continue. As mentioned above, you can install and run Windows for an unlimited time without activating.

Setup will present to you a list of Windows versions. Windows 11 Home or Windows 11 Pro are the valid choices. Windows 11 Home is the standard version that includes all the main features for most users. Windows 11 Pro offers all the features of Home but adds BitLocker encryption, remote desktop, and additional tools for business use, like policy management. Unless you already know that you need the Pro features, go with Windows 11 Home.

Select custom install when the option appears, then setup will ask where you want to install Windows. This is when having only one drive installed comes in handy. Windows setup does not provide detailed information about the interfaces where your drives are connected, when you have multiple identical drives. Also, if only one drive is available, Windows will place its bootloader on the drive where the operating system is being installed. However, if multiple drives are connected, Windows *might* install the bootloader on a different drive, which can cause issues if that drive is removed or modified later on. Select the unallocated space on the drive and setup will automatically create a partition for the operating system.

After a restart, setup will gather information from you about the region, country and keyboard layout, then you'll be prompted to follow the instructions to connect your network.

One of the last steps will be configuring the initial privacy settings. Look through them and decide which options you want to enable. Since you've built a desktop PC, the **Find my device** option doesn't really apply, so that can certainly be disabled.

4.3 Running Windows Update
After installing Windows, your first priority is to ensure your system is running the latest updates. Running **Windows Update** checks for all essential system updates and, if needed, automatically applies them to your PC.

To check for updates, open the Start menu, search for Windows Update and select **Check for updates**. Install any available updates. Repeat this process until no further updates are found. In the future, Windows will prompt you when additional updates are ready.

Under Advanced Options in Windows Update, you'll find several settings to customize. I recommend enabling **Receive updates for other Microsoft products** if you use Microsoft Office or similar software, so all updates are managed together. To avoid unexpected restarts, turn on **Notify me when a restart is required** to finish updating and set your **Active hours** to prevent work interruptions.

4.4 Installing Drivers for Components
Windows provides drivers for most hardware, but some might be missing. To check, open **Device Manager** (search for it in the taskbar) and look for any devices marked with a yellow exclamation mark. Download the driver files from your motherboard's support page (and GPU, if applicable). After downloading, extract any .zip files and run the .exe to install the drivers. Once the drivers are installed, and after restarting if needed, go back into the **Device Manager** panel and check to see that all of your devices are recognized. You can also right-click the device and select **Update driver**, but it may not be able to locate the correct files.

While you are looking for the driver files, you'll probably see some other "control center" or "experience" type software for download. As a rule, these aren't needed, and can clutter up your system. For example, Gigabyte offers their Control Center application. It installs several small applications for controlling various functions that are of marginal usefulness, while not installing any needed driver files. Similarly, NVIDIA offers drivers for their GPU cards, as well as their "GeForce Experience" software, which is not necessary for complete functionality of the GPU card. Control programs like these are similar to the "bloatware" that comes with many prebuilt computers. They mostly just take up space and system resources.

4.5 Privacy & Security Settings
By default, Windows collects some data with permissions that you might want to adjust based on your preferences. Navigate to **Privacy & Security > General** from the **Settings** menu and review the options. While some should be left on, others, such as data-sharing settings, can be adjusted for more privacy. I leave on **Let Windows improve Start and search results** and **Show me notifications in the settings app**, and turn off other data collection options.

Back under the main Privacy & security panel you'll see three sections: Security, Windows permissions, and App permissions. Review these and set them to your preference.

Search **Windows security** in the task bar and review the available options under Security at a glance. You'll notice that **Windows comes with virus & threat protection, and a firewall,** which are both enabled by default. These are as good as, or better than, third party software and I highly recommend leaving them enabled.

4.6 Backup, Restore, & Recovery

Windows offers several built-in tools to help you preserve your data and protect your system. These include using Backup and Restore (Windows 7), setting system restore points, and creating a USB recovery drive. In the following sections, I'll explain how to use each of these features.

Using Backup and Restore (Windows 7): This feature is still available in Windows 11. Despite its name, this tool remains a reliable way to create backups of your personal files and create a system image of your entire computer.

To set up a backup using this tool, click on the Start menu, type **Control Panel** in the search bar, and select it from the results. Once in the Control Panel, navigate to **System and Security**, and then select **Backup and Restore (Windows 7)** from the list.

Next, click on "Set up backup" on the right side of the window. You'll then be prompted to select where you want to save your backup. If you have an HDD installed for backups, select that drive as the destination, or any other location you choose, and click "Next."

You will have the option to let Windows choose what to back up or to select the files yourself. Choosing "Let Windows choose" is recommended, as it will back up data files saved in libraries, on the desktop, and in default Windows folders. Alternatively, you can select "Let me choose" to manually pick the folders and files you want to back up. Select this option if you've created your own directory structure for saving files outside of the default Windows "My Documents" folder (highly recommended). There is also a checkbox **Include a system image** which will save a copy of the drive required for Windows to run which can be used to restore your computer. After making your selection, click "Next."

A side note on saving your files: I have a system for saving my files that I recommend you adopt. While you can use a single storage device, I prefer multiple M.2 NVMe drives. I name one drive something short and clear, like "Data" or "Files," and that's where I store all my work. On this drive, I create a directory of folders to organize my files and projects, and I back it up regularly. I avoid using the default Windows folders, "Documents," "Pictures," "Videos", because Windows handles those folders differently, particularly with regards to its backup and sync service, OneDrive. Creating your own folders lets you maintain full control over your files.

Review your backup settings to ensure everything is correct. If you wish to adjust how often backups occur, click "Change schedule." Once you're satisfied with the settings, click "Save settings and run backup." Windows will begin the backup process and will continue to back up your files according to the schedule you've set. Regular backups will help you prevent data loss due to hardware failure, accidental deletion, or malware.

Setting System Restore Points: This allows you to roll back your system files and settings to a previous state without affecting your personal files. This feature can be useful if something goes wrong after installing new software, drivers, or updates.

To create a system restore point, access the System Protection settings by clicking on the Start menu, typing **Create a restore point** in the search bar, and selecting it from the results. This action opens the System Properties window on the System Protection tab.

Under Protection Settings, you'll see a list of available drives. Ensure that protection is turned on for your system drive labeled "C:". If it's not enabled, select the drive and click "Configure." Choose "Turn on system protection," adjust the Max Usage slider to allocate disk space for restore points, and click "OK" to save the settings.

In the System Protection tab, click the "Create" button to make a new restore point. Enter a description for the restore point, such as "Before installing graphics driver update," and click "Create." Windows will create the restore point, which may take a few minutes. A confirmation message will appear once it's completed. It's a good idea to create a restore point before making major changes to your system. Windows also automatically creates restore points during system updates and other events.

Creating a USB Recovery Drive: A USB recovery drive can help you troubleshoot and restore your system if it fails to start or encounters serious issues. This bootable drive gives you access to advanced recovery options.

To create a USB recovery drive, you'll need a USB flash drive with at least 16 GB of storage, which will be erased. Insert the USB flash drive into your computer. Next, open the Recovery Drive tool by clicking on the Start menu, typing **Create a recovery drive** in the search bar, and selecting it from the results. In the Recovery Drive window, you'll see an option labeled **Back up system files to the recovery drive**. It's recommended to keep this option checked, as it allows you to reinstall Windows from the recovery drive. Click "Next" to continue.

The tool will scan for available USB drives. Select your USB flash drive from the list and click "Next." A warning will appear stating that everything on the drive will be deleted. Click "Create" to begin the process. The process may take an hour or more, especially if you're including system files. Once the recovery drive is ready, you'll see a final confirmation screen. Click "Finish" to exit.

Keep the recovery drive in a safe place. In case your computer experiences major issues, you can boot from this USB drive to access recovery tools or reinstall Windows if necessary.

Using these built-in Windows features, you'll have multiple layers of protection for your data and system configuration. Regularly backing up your files, maintaining system restore points, and having a recovery drive on hand will help you quickly recover with minimal disruption.

4.7 Disk Management

At this point in your build, you probably have one M.2 NVMe SSD installed at the M.2 slot closest to the CPU on your motherboard. You may have purchased multiple drives to store your data. To add another drive after Windows is installed, the Windows Disk Management utility comes into play. Here I'll go over what happens when Windows recognizes you've added a new drive to your system, and what to do when Windows doesn't automatically pick up the new drive.

Adding a Drive: After you get the new drive installed in the PC case, boot the computer and enter the BIOS screen during boot to confirm that the drive has been correctly recognized by your motherboard.

When you install a new drive in your PC, Windows will usually recognize it automatically. In this case, you'll receive a notification saying that a new drive has been found and requires initialization. Windows will guide you through this process.

To start, Windows may prompt you to initialize the drive. Here, you'll be asked to choose a partition style, GPT (GUID Partition Table) or MBR (Master Boot Record). For your Non Gaming PC build, GPT is the recommended choice, as it offers more flexibility and supports larger drives.

Once the partition style is selected, the drive will need to be formatted before it can be used for storing data. Windows will guide you through formatting the drive to the NTFS file system, which is the default. During this process, you can also assign a drive letter to make it easier to access in File Explorer (e.g., "D:" or "E:"). For this build, the existing drive is C:, so make your new drive the D: drive. (Note: Don't skip over drive letters, because sometimes this can cause unexpected behavior later down the line with removable drives, network drives, scripts, or older applications.) Once formatted, the drive will be ready for use.

If you don't see a pop-up or notification when the drive is first installed, don't worry. It may still be recognized, and you can go ahead with manual initialization, as described in the next section.

How to Manually Initialize a Drive in Windows: If Windows doesn't automatically recognize your new drive, you'll need to manually initialize it using the Disk Management tool. To do this, right-click on the Start button (or press `Win + X` on your keyboard) and select **Disk Management** from the menu. This tool provides an overview of all the storage drives connected to your system. In Disk Management, look for the new drive, which may appear as "Unknown" or "Not Initialized." It will likely be shown as unallocated space, indicating that it hasn't been set up for use yet. Right-click on the drive name (listed on the left-hand side of the entry) and select **Initialize Disk.** You'll then be prompted to choose a partition style, GPT or MBR. As mentioned earlier, GPT is generally recommended. Once you've selected the partition style, click OK to continue.

After initialization, the drive still needs to be formatted. Right-click on the unallocated space of the newly initialized drive and select New Simple Volume. The wizard that appears will guide you through setting the drive size, assigning a drive letter, and choosing the NTFS file system. Once these steps are completed, the drive will be formatted and ready to use.

If the new drive doesn't appear in Disk Management, there could be an issue with the drive's connection to the system. Check to ensure the drive is properly connected and verify that the drive is recognized in the system BIOS.

4.8 System Performance

Now that Windows is installed I'll discuss enabling XMP, optimizing Windows performance, and customizing the Windows interface for your Non Gaming PC.

Enable XMP in your BIOS: Restart the computer and enter the BIOS using the keystroke shown during boot (Delete, F2, F12...). In the BIOS, look for Extreme Memory Profile (X.M.P). In GIGABYTE's BIOS, it's on the front page. Set it to XMP1 or similar, and check that the memory frequency matches the advertised frequency of your memory modules. Save and exit. On reboot, the system may spend some time doing memory training, where the motherboard is testing and configuring the RAM to confirm it operates correctly at the specified settings. If you've selected memory modules from the qualified vendors list (QVL) for your motherboard, there shouldn't be any problems enabling XMP. When the system comes back online, enter the BIOS and confirm that the memory speed is correct. For a detailed look at your hardware, I recommend installing CPU-Z from www.cpuid.com to see details about your memory, motherboard and CPU.

Optimize Windows System Performance: Adjusting Power Options, Startup Programs and the appearance and performance of Windows are easy ways to get the most out of your build. To change power options, open **Control Panel > Hardware and Sound > Power Options**, where you can choose or customize a power plan. High Performance is the best desktop PC setting. Startup programs can be controlled in Task Manager. Right-click the taskbar, select Task Manager, then click Startup Apps on the right edge of the window to get a list of applications that start with Windows. The status column shows if the app is enabled or disabled. Right-click any of the apps you aren't familiar with to get more information or run an online search to find out more. If you're sure you don't need it, right-click and select **Disable**, and the next time you start Windows, that app won't run. Unneeded apps running at startup can increase Windows boot time and drag down your system, so it's not a bad idea to check this list once in a while.

The special effects in the Windows interface, like drop shadows and animations, can eat up valuable system resources. To adjust what effects you want to see, click the Start button and type **View advanced system settings**, then click the **settings button** under Performance. On the Performance Options panel's Visual Effects tab, there are 17 checkboxes to control animations, shadows, and other special effects in the operating system. At the top you'll see 4 options: Let Windows Choose, Best Appearance, Best Performance, and Custom. Best performance turns all 17 options off, but leaves the OS feeling a little flat. I like to run **Let Windows Choose**, which leaves nearly everything on. Some of the more worthwhile options are Smooth edges of screen fonts, Show window contents while dragging, and Use drop shadows for icon labels on the desktop. Play around with these and adjust them to your liking.

Customizing the Windows interface: Tailor the Windows environment to suit your needs with the following adjustments.

In File Explorer, you can set the view to your preference in one folder, then set all folders in the system to have the same view. To do this, open File explorer by clicking the yellow folder icon in the taskbar. Navigate to a folder with many files and documents. Click the view menu at the top of the window and select Details or any other view you prefer. After everything looks the way you want, click the 3 dots next to View, and select Options at the bottom. In the Folder Options panel, select the View tab. In the Folder Views section at the top, click the **Apply to Folders** button and select Yes to apply the current folder view to all similar folders in the system.

The Taskbar has some useful settings. Right-click the Taskbar to get to settings or launch Task Manager. In Taskbar settings you can change the search to an icon only, icon and label, or a search box. I prefer the icon only as it takes up less room. Another option in the taskbar is Widgets, which adds an icon to display an endless scrolling feed of headlines, weather, games and some ads. I leave Widgets turned off. Taskview plants a button on the taskbar that lets you zoom out to display all open application windows, which can be useful.

The Start menu has a few little tricks. Right-click the Start icon to get the Quick Link menu, which has many system related options like Disk Management, Task Manager, Settings, and more. Click the Start icon and then right-click anywhere on the Start menu and select Start Settings to open the Personalization > Start panel. It lets you control how many apps you see in the Start Menu and other things like recommendations, recently added apps, and you can select links to add next to the power button.

The desktop also has options. Right-click the Start icon and select desktop from the Quick Link menu to quickly clear the screen and show your desktop. Right-click the desktop and select Display Settings to access the setup options for your display including multiple monitors, scaling (100% is best), resolution, and others. Right-click the desktop and choose Personalize to see options for background, colors, themes, and other customization.

In this brief Windows Guide, you've learned how to install and configure Windows, update your system, manage drivers, adjust privacy and security settings, and create backups for data protection. By following these steps, you ensure that your Non Gaming PC is set up securely, performs optimally, and remains easy to maintain. From enabling essential updates to customizing your interface and safeguarding your data, these foundational practices will help you get the most out of your Windows experience and keep your system running smoothly.

www.ingramcontent.com/pod-product-compliance
Lightning Source LLC
Chambersburg PA
CBHW060452060326
40689CB00020B/4506